IV International Scientific Conference on Advanced Technologies in Aerospace, Mechanical and Automation Engineering (MIST Aerospace-IV 2021)

IOP Conference Series: Materials Science and Engineering
Volume 1227

Krasnoyarsk, Russia
10 – 11 December 2021

Editors:

Igor V. Kovalev
Alena A. Stupina
Anna A. Voroshilova

ISBN: 978-1-7138-4685-7
ISSN: 1757-8981

Printed from e-media with permission by:

Curran Associates, Inc.
57 Morehouse Lane
Red Hook, NY 12571

CURRAN ASSOCIATES INC. proceedings.com

Some format issues inherent in the e-media version may also appear in this print version.

Institute of Physics
Dirac House, Temple Back
Bristol BS1 6BE UK

Phone: 44 1 17 929 7481
Fax: 44 1 17 920 0979

techtracking@iop.org

Additional copies of this publication are available from:

Curran Associates, Inc.
57 Morehouse Lane
Red Hook, NY 12571 USA
Phone: 845-758-0400
Fax: 845-758-2633
Email: curran@proceedings.com
Web: www.proceedings.com

TABLE OF CONTENTS

IOP Conf. Series: Materials Science and Engineering 1227 (2022) 011001 doi:10.1088/1757-899X/1227/1/011001

Overview of IV International Conference on Advanced Technologies in Aerospace, Mechanical and Automation Engineering – MIST Aerospace-IV-2021

I V Kovalev[1,2,3,4], N A Testoyedov[4,5] and A A Voroshilova[2]

[1] Siberian Federal University, 79, Svobodny pr., Krasnoyarsk, 660041, Russia
[2] Krasnoyarsk Science and Technology City Hall, 61, Uritskogo Str., Krasnoyarsk, 660049, Russia
[3] China Aviation Industry General Aircraft Zhejiang Institute Co., Ltd, China
[4] Reshetnev Siberian State University of Science and Technology, 31, Krasnoyarsky Rabochy Av., Krasnoyarsk, 660037, Russia
[5] JSC "Academician M F Reshetnev Information satellite systems", 52 Lenin Street, Zheleznogorsk, Krasnoyarsk region, 662972, Russia

E-mail: krasnio@bk.ru

Abstract. The overview describes the main directions and results of the IV International Conference on Advanced Technologies in Aerospace, Mechanical and Automation Engineering (MIST Aerospace-IV-2021) held in Krasnoyarsk on 10-11 December 2021. It gives the details about the participants and the proceedings. The purpose of the Conference is to share the experience of leading experts in the application of advanced science-intensive and information technologies in the aerospace industry, mechanical engineering and automation of technological processes and production.

We are pleased to present this issue of IOP Conference Series: Materials Science and Engineering (MSE) for the selected revised papers presented in the IV International Conference on Advanced Technologies in Aerospace, Mechanical and Automation Engineering (MIST Aerospace-IV-2021) on 10-11 December 2021 at the Krasnoyarsk Science and Technology City Hall (Russia). The purpose of the Conference is to share the experience of leading experts in the application of advanced science-intensive and information technologies in the aerospace industry, mechanical engineering and automation of technological processes and production.

MIST Aerospace-IV-2021 was jointly organised by the International and Russian Union of Scientific and Engineering Associations (Moscow, Russia), School of Space and Information Technology of the Siberian Federal University and Krasnoyarsk Science and Technology City Hall.

The main partner organisations in preparing and organising the Conference in Krasnoyarsk as well as of the previous scientific events [1-16] were: JSC "Academician M F Reshetnev Information Satellite Systems" and Krasnoyarsk Branch of the Russian Engineering Union.

The papers of this volume are related to the following themes:

- Aerospace Technologies and Industrial Production;
- Reliability and Data Protection in Automation Systems;

- Cybernetics, Automation and Information Technologies;
- Energy, Chemical and Ecological Engineering.

The Program Committee and Editorial Board included prominent professors and scientists from the Russian Academy of Sciences, Amity University (India), Republican Centre for Space Communications and Electromagnetic Compatibility of Radio-Electronic Devices (Kazakhstan), Bukhara Engineering Technological Institute and Tashkent Institute of Irrigation and Agricultural Mechanization Engineers (Uzbekistan), Siberian Federal University (Russia), Bialystok University of Technology and West Pomeranian University of Technology (Poland) as well as specialists from Russian and foreign leading industrial enterprises.

The event has offered a platform for bringing together students, postdocs, innovative academics and industrial experts to exchange their ideas and contribute new engineering approaches to research technological and automation processes in various production systems. Great interest was aroused by the section about material science and innovative technology and the section about energy, chemical technologies and ecological engineering.

The participants from Russia represented universities, scientific institutes and industrial enterprises from more than 15 regions:

- Bauman Moscow State Technical University, Moscow, Russian Federation
- Bryansk Technical State University, Bryansk, Russian Federation
- K.G. Razumovsky Moscow State University of technologies and management (the First Cossack University), Moscow, Russian Federation
- Kadyrov Chechen State University, Chechen Republic, Russian Federation
- Kuban State Technological University, Krasnodar, Russian Federation
- Kuzbass State Agricultural University, Kemerovo, Russian Federation
- Novosibirsk State Technical University (NSTU), Novosibirsk, Russian Federation
- Orenburg State University, Orenburg, Russian Federation
- Penza State Technological University, Penza, Russian Federation
- Platov South-Russian State Polytechnic University (NPI), Novocherkassk, Russian Federation
- Plekhanov Russian University of Economics, Moscow, Russian Federation;
- Saint Petersburg State University of Aerospace Instrumentation, SPB, Russian Federation
- Saint Petersburg University of the ministry of Internal Affairs of Russia, SPB, Russian Federation
- Samara National Research University, Samara, Russian Federation
- Samara University, Samara, Russian Federation
- Vladimir State University, Vladimir, Russian Federation
- Vyatka State University, Kirov, Russia

The Conference provided the premier interdisciplinary and multidisciplinary forum for researchers, practitioners and educators to present and discuss the most recent advanced science-intensive and information technologies in the aerospace industry, mechanical engineering and automation of technological processes and production.

All participants were invited to present their papers in this Volume and all submitted manuscripts went through the independent peer review process. We are very grateful to all reviewers from Russia, Kazakhstan and Uzbekistan for their time and highly professional comments. We deeply believe that their reviews gave opportunity to improve the scientific quality of the presented papers which may be useful for academic, scientific and industrial partners.

Each paper has been peer reviewed by the reviewers specialized in the related field. All papers that have been selected and accepted for publication reflect modern engineering approaches in all

Conference directions, broaden the researches of the previous MIST Aerospace Conferences [1,5,11]. We hope that you enjoy the reading of the papers.

We chose the IOP Conference Series: Materials Science and Engineering to provide all contributors with the opportunity to publish their papers in an international, peer-reviewed journal. This is understood and appreciated by all the participants of our Conference, and therefore this Volume provides an excellent overview of the main topics of our Conference.

Acknowledgements
The JSC "Academician M F Reshetnev Information Satellite Systems" and Siberian Scientific Centre DNIT sponsored the Conference.

MIST Aerospace-IV 2021 Conference Committee highly appreciate the contribution of Professor Nikolay Testoyedov, corresponding member of the Russian Academy of Sciences, Director General of Information Satellite Systems – Reshetnev Company (Zheleznogorsk, Krasnoyarsk region, Russia) for cooperation in organization of MIST Aerospace-IV 2021 and preparation of the papers for this Volume.

We are grateful to scientists and professors of Siberian Federal University, Saint-Petersburg State University of Aerospace Instrumentation, Samara National Research University and Samara University for active participation in the Conference.

On behalf of the Conference Committee and organizers, we would like to thank all the authors who contributed to this Volume as well as to the reviewers, speakers and all the conference participants for their support to MIST Aerospace-IV 2021.

We express gratitude to IOP Publishing for an opportunity to publish the Proceedings of the Conference to provide open access and to make them available for worldwide recognition.

References

[1] 2018 *IOP Conf. Ser.: Mater. Sci. Eng.* **450** 011001
[2] Kovalev I V, *et al.* 2019 *IOP Conf. Ser.: Earth and Environ. Sci.* **315** 011001
[3] Kovalev I V, *et al.* 2019 *IOP Conf. Ser.: Mater. Sci. Eng.* **537** 011001
[4] Kovalev I V, *et al.* 2019 *J. Phys.: Conf. Ser.* **1399** 011001
[5] Kovalev I V and Voroshilova A A 2020 *IOP Conf. Ser.: Mater. Sci. Eng.* **734** 011001
[6] Kovalev I V, Pyzhikova N V and Voroshilova A A 2020 *IOP Conf. Ser.: Earth and Environ. Sci.* **421** 011001
[7] Antokhina Yu A, *et al.* 2020 *J. Phys.: Conf. Ser.* **1515** 011001
[8] Kovalev I V and Voroshilova A A 2020 *IOP Conf. Ser.: Mater. Sci. Eng.* **862** 011001
[9] Kovalev I V, *et al.* 2020 *IOP Conf. Ser.: Mater. Sci. Eng.* **919** 011001
[10] Kovalev Igor V, *et al* 2020 *J. Phys.: Conf. Ser.* **1679** 011001
[11] Kovalev Igor V, *et al* 2021 *IOP Conf. Ser.: Mater. Sci. Eng.* **1047** 011001
[12] Antokhina Yu A, *et al* 2021 *J. Phys.: Conf. Ser.* **1889** 011001
[13] Kovalev Igor V, *et al* 2021 *IOP Conf. Ser.: Mater. Sci. Eng.* **1155** 011001
[14] Eugene Semenkin, Igor Kovalev (eds.) CEUR-WS.org **2899** https://doi.org/10.47813/dnit-mip3/2021-2899
[15] Kovalev I V *et al* 2021 *IOP Conf. Ser.: Mater. Sci. Eng.* **1181** 011001
[16] Kovalev I V and Testoyedov N A 2021 *AIP Conference Proceedings* **2402**, 010001

International Programme Committee

Co-Chairperson Yury V. Gulyaev - Academician of the Russian Academy of Science, Doctor of Physics and Mathematics, Professor, President of the International and Russian Union of Scientific and Engineering Associations, Moscow, Russia

Co-Chairperson Nikolay Testoyedov - Corresponding member of the Russian Academy of Sciences, Doctor of Engineering, professor of the Satellites Department at Reshetnev University, General Director of ISS Reshetnev Company, Krasnoyarsk, Russia

Vice-Chairman Viktor Lefter – Doctor of Technical Sciences, Professor, Scientific Advisor of the jointstock company "Republican center for space communications and electromagnetic compatibility of radioelectronic devices», Nur-Sultan, Kazakhstan

Vice-Chairman Igor V. Kovalev - Doctor of Technical Sciences, Professor, President of the Krasnoyarsk Regional Union of Scientific and Engineering Associations, Director of Krasnoyarsk Regional Science and Technology City Hall, Krasnoyarsk, Russia

Nusratilla R. Barakaev - Doctor of Technical Sciences, Professor, Rector of Bukhara Engineering Technological Institute Bukhara, Uzbekistan

Odilzhon Mamatkarimov - Doctor of Physics and Mathematics, Professor, Rector of Namangan Engineering Technological Institute Namangan, Uzbekistan

Vasiliy F. Shabanov - Academician of the Russian Academy of Science, Doctor of Physics and Mathematics, Professor, Scientific Supervisor of the Krasnoyarsk Science Centre of the Siberian Branch of the Russian Academy of Science Krasnoyarsk, Russia

Sergey N. Varnakov - Doctor of Physical and Mathematical Sciences, Professor, Vice-Director of Krasnoyarsk Science Center of the Siberian Branch of the Russian Academy of Sciences Krasnoyarsk, Russia

Takhir Z. Sultanov - Doctor of Technical Sciences, Professor, Vice-Rector for Research and Innovation of the Tashkent Institute of Irrigation and Agricultural Mechanization Engineers Tashkent, Uzbekistan

Mirzo Z. Sharipov - Doctor of Physics and Mathematics, Professor, Vice-rector for research and innovation of Bukhara Engineering Technological Institute Bukhara, Uzbekistan

Sergey P. Drukarenko - Vice-president, First Secretary the International and Russian Union of Scientific and Engineering Associations Moscow, Russia

Oybek Ergashev - Doctor of Chemistry, Professor, Vice-Rector for Science and Innovations of Namangan Engineering Technological Institute Namangan, Uzbekistan

Dr Krystyna Araszkiewicz - West Pomeranian University of Technology Szczecin, Poland

Krishna Mohan Agarwal - Assistant Professor, Grade III, Mechanical Engineering Department, Amity University Uttar Pradesh Noida, India

Bakhytbek Zhankubaev - Serpin Consult Group Almaty, Republic of Kazakhstan

Satya N Mandal - RICS School of Built Environment, Amity University Noida, India

Adam Musiuk - Bialystok University of Technology, Poland

Bahadir S. Mirzaev - Doctor of Technical Sciences, Professor, First Vice-Rector of the Tashkent Institute of Irrigation and Agricultural Mechanization Engineers Tashkent, Uzbekistan

Kadyrzhan A. Shovazov - Candidate of Technical Sciences, Associate Professor, Dean of the Faculty of Agricultural Mechanization of the Tashkent Institute of Irrigation and Agricultural Mechanization Engineers Tashkent, Uzbekistan

Avazbek Obidov - Doctor of Technical Sciences, Professor, Dean of the Faculty of Technology-Engineering, Namangan Engineering Technological Institute Namangan, Uzbekistan

Khamro B. Dustov - Editor-in-chief of the journal "Development of Science and Technology", Doctor of Chemistry, Professor of the Department of Oil and Gas Engineering, Bukhara Engineering Technological Institute Bukhara, Uzbekistan

Nosir Sharibaev - Doctor of Physics and Mathematics, Professor, Head of the Department: "Energy", Namangan Engineering Technological Institute, Conference coordinator Namangan, Uzbekistan

Ismoil B. Isabaev - Doctor of Technical Sciences, Professor of the Department of Food Technology, Bukhara Engineering Technological Institute Bukhara, Uzbekistan

Sadriddin F. Fozilov - Doctor of Technical Sciences, Professor, Head of the Department "Technology of Chemical Gas Processing", Bukhara Engineering Technological Institute Bukhara, Uzbekistan

Komil D. Astanakulov - Doctor of Technical Sciences, Senior Researcher, Head of the Department of Agricultural Machinery, Tashkent Institute of Irrigation and Agricultural Mechanization Engineers Tashkent, Uzbekistan

Hasan I. Turkmenov - Candidate of Technical Sciences, Associate Professor, Head of the department of "General technical disciplines" of the Tashkent Institute of Irrigation and Agricultural Mechanization Engineers Tashkent, Uzbekistan

Solikh K. Astanov - Doctor of Physical and Mathematical Sciences, Professor, Bukhara Engineering Technological Institute Bukhara, Uzbekistan

Khairullo F. Dzhuraev - Doctor of Technical Sciences, Associate Professor, Dean of the Faculty of Technological Process Control Systems, Bukhara Engineering Technological Institute Bukhara, Uzbekistan

Mubin M. Vakhitov - Doctor of Technical Sciences, Professor of the Department of Architecture, Bukhara Engineering Technological Institute Bukhara, Uzbekistan

Hamid Akhmedkhodjaev - Doctor of Technical Sciences, Professor of the Professor of the Department "Technology of processing of natural fibers", Namangan Engineering

Peer review declaration

IV International Conference on Advanced Technologies in Aerospace, Mechanical and Automation Engineering (MIST Aerospace-IV-2021), 10-11 December 2021, Krasnoyarsk, Russian Federation

All papers published in this volume of IOP Conference Series: Materials Science and Engineering have been peer reviewed through processes administered by the Editors. Reviews were conducted by expert referees to the professional and scientific standards expected of a proceedings journal published by IOP Publishing.

- **Type of peer review: Double-blind**

All papers came through the basic review which included an initial technical criteria check (paper field, structure of submission, adherence to the submission instructions, English language usage and a check for the similarity rate). Any papers out of the scope or containing plagiarism, including self-plagiarism, were rejected.

We used a double-blind system for peer review; both reviewers' and authors' identities remained anonymous. The submitted papers were reviewed by at least two experts: one editorial staff member as well as at least one external reviewer. The third/fourth reviewer is involved in case two reviewers have disputes about the content of the papers or the authors do not agree with the review result. The review process took from 7 to 15 days as a rule. The reviews were conducted to the professional and scientific standards.

The decision to accept or reject the paper was based on the suggestions of reviewers. Acceptance/rejecting notifications were sent to the corresponding author(s).

- **Conference submission management system:**

Conference submission is carried out through the conference management system which was developed by the Organizer. It easy to use, and has features to make it suitable for simultaneous conference calls.

More than 20,500 registration forms have been already processed by the system which showed it stability and flexibility.

The conference management system is integrated with the conference website of the Organizer which helps the Organizer to cope with the complexity of the submission, reviewing, refereeing and accepting processes. The current version supports:

1. management of the access of reviewers and editors to papers;
2. automatic registration for conferences and paper submission;
3. list of the current, forthcoming and held conferences with all necessary information;
4. list of all conference materials with online presentations and/or video reports of participants, list of published proceedings with direct links to publications and journals;
5. sending emails to programme committee members, reviewers and authors;
6. preparation and publishing of conference programmes;

7. all authors submitting the papers are assumed to accept the terms of Publication Licence;
8. all requirements for papers, templates and sample papers are accessible for the authors.

- **Number of submissions received:** 29
- **Number of submissions sent for review:** 29
- **Number of submissions accepted:** 12
- **Acceptance Rate (Number of Submissions Accepted / Number of Submissions Received X 100):** 41
- **Average number of reviews per paper:** 2
- **Total number of reviewers involved:** 12
- **Any additional info on review process:**

The Checklist for the reviewers included the following criteria: relevance to the scope of the conference topics, the scientific quality, assessing the errors, organisation of the paper, the title, the references, the level of the English language, the quality of the figures and tables, etc.

The reviewers could give the following recommendation: to publish the paper as it is, to publish it after mandatory/optional minor/major revision, to reject the paper.

- **Contact person for queries:**

Anna Voroshilova, Krasnoyarsk Science and Technology City Hall, 61, Uritskogo Street, Krasnoyarsk, 660049, Russia, E-mail: krasnio@bk.ru. Tel + 7 913 538 6990

Resistance law dependence on radius ratios for turbulent Taylor-Couette flow from extended local balance model

Alexander Balonishnikov[1,3] and Julia Kruchkova[2]

[1]Department of Mathematics and Mechanics, Department of Physics, Saint-Petersburg State University of Aerospace Instrumentation, 67, Bolshaya Morskaya str., SPb 190000, Russia
[2]Department of foreign Languages, Saint Petersburg University of the ministry of Internal Affairs of Russia, 1 Pilot Pilyotov str., SPb 198206, Russia

[3]E-mail: balonishnikov@mail.ru

Abstract. Turbulent Couette-Taylor flow is the most interesting kind of flow for theoreticians and experimenters. There are a great number of theories and experiments for larger and larger Reynolds numbers. The picture becomes more complex if we consider the case of independently rotating cylinders. The most simple quantity is a torque, acting on the lateral surface of inner cylinder by rotating fluid. How can we deduce theoretically the torque from Navier- Stokes equation? Unfortunately we have only bounds on this variable yet. So we cause use models of turbulence to predict the torque dependence on Reynolds numbers or Taylor number and on radius ration. Here we present the results of simple calculations, based on extended local balance model. The study compares the theory with experiments.

1. Introduction

Turbulence remains the most difficult unsolved problem of classical physics. A profound review of the topic has been recently published in [1]. The basis of achieved knowledge are Navier – Stokes equation as a starting point of research, as well as Kolmogorov – Richardson picture of energy cascade from large eddies to smaller ones, Prandl – von Karman law of wall. There are many turbulence models, but scientists still have not reached an agreement about which model to consider the best. The most simple kinds of turbulent flow are flows in direct round pipe and between rotating coaxial rotating cylinders (Taylor – Couette flow). Two close theoretical approaches to turbulent Taylor – Couette flow have been recently published in [2, 3]. A concurrent approach to this problem, based on the extended local balance model [4-8], is being presented here.

2. Formulas for the analysis.

The dependence of Nusselt number (Nu) on the cylinder radii is considered. It is assumed that only inner cylinder rotates. Following [9],

$$Nu = \frac{G}{G_{lam}} , \tag{1}$$

where $G = \dfrac{T}{\rho v^2 L} = 2\pi a^2 u_{*1}^2 / v^2$ is turbulent dimensionless torque, $T-$ dimensional torque, $u_{*1}-$ is the friction velocity at the lateral surface of the inner cylinder, $\rho-$ is density of fluid, $\nu-$ molecular viscosity, $L-$ is the length of the cylinders,

$a-$ is the radius of inner cylinder $G_{lam}-$ is laminar dimensionless torque. The expression for G_{lam}, considering both cylinders independent rotations, can be find in [10]:
$G_{lam} = 4\pi v^{-1} a^2 b^2 |\omega_1 - \omega_2| (b^2 - a^2)^{-1}$, where ω_1- is the angular velocity of the inner cylinder, ω_2- is the angular velocity of the outer cylinder, $b-$ is the radius of the outer cylinder.
Here we multiplied the expression [10] by 2π, because some authors prefer definition $G = \dfrac{T}{2\pi\rho v^2 L}$ [10]. Nusselt number will be the same. We have the case, when the inner cylinder rotates and the outer cylinder is at rest:

$$G_{lam} = \frac{4\pi\eta \operatorname{Re}_1}{(1+\eta)(1-\eta)^2}, \tag{2}$$

the radius ratio $\eta = 1/q = a/b$, $b-$ is the radius of outer cylinder, Reynolds number $\operatorname{Re}_1 = \omega_1 a(b-a)/\nu$, ω_1- angular velocity of inner cylinder, ω_2 -angular velocity of outer cylinder, $\omega_2 = 0$,
$\nu-$ is the fluid molecular viscosity. The correspondence of Reynolds number and Taylor number Ta can be found in [2]:

$$Ta = \frac{(1+\eta)^6}{64\eta^4} \operatorname{Re}_1, \tag{3}$$

where Taylor number is $Ta = \dfrac{1}{4} \dfrac{\sigma (b-a)^2 (b+a)^2 (\omega_2 - \omega_1)^2}{v^2}$, pseudo-Prandtl number $\sigma = \dfrac{(1+\eta)^4}{(4\eta)^2}$.

We used the expression for G [4]:

$$G = 2\pi d^2 / [(W(de^c)]^2 \tag{4}$$

where $c = \{4\pi q J_0(q)/(q^4 - 1) + q \ln[C(q-1)] + q^{-1} \ln[C(q-1)(q-1)]\}/(q + q^{-1})$,

$C = 10.0, \kappa = 0.4$ - are the Prandtl- von Karman's constants, $d = \dfrac{8\kappa q^3 \sqrt{Ta}}{(q^4 - 1)(q+1)^2}$,

$$J_0 = \int_1^q \{x \csc[\pi(x^4 - 1)/(q^4 - 1)] - (q^4 - 1)(4\pi)^{-1}/(x-1) - (q^4 - 1)(4\pi)^{-1} q^{-2}/(q-x)\} dx,$$

W(x) – is the Lambert function.
The Prandl- von Karman's constants include in the well-known wall functions:

$U(r)=\omega_1 a-\frac{u_{*1}}{\kappa}\ln\frac{Cu_{*1}(r-a)}{\nu}$, $U(r)=\frac{u_{*2}}{\kappa}\ln\frac{Cu_{*2}(b-r)}{\nu}$, r- is the radial coordinate, $U(r)-$ is the mean velocity, $u_{*2}-$ the friction velocity on the lateral surface of the outer cylinder.

3. Comparison with experiments and discussion

The dependences of the Nusselt number (Nu) on Taylor number (Ta), using the formulas (1), (4) and experiments, is given on Figure 1.

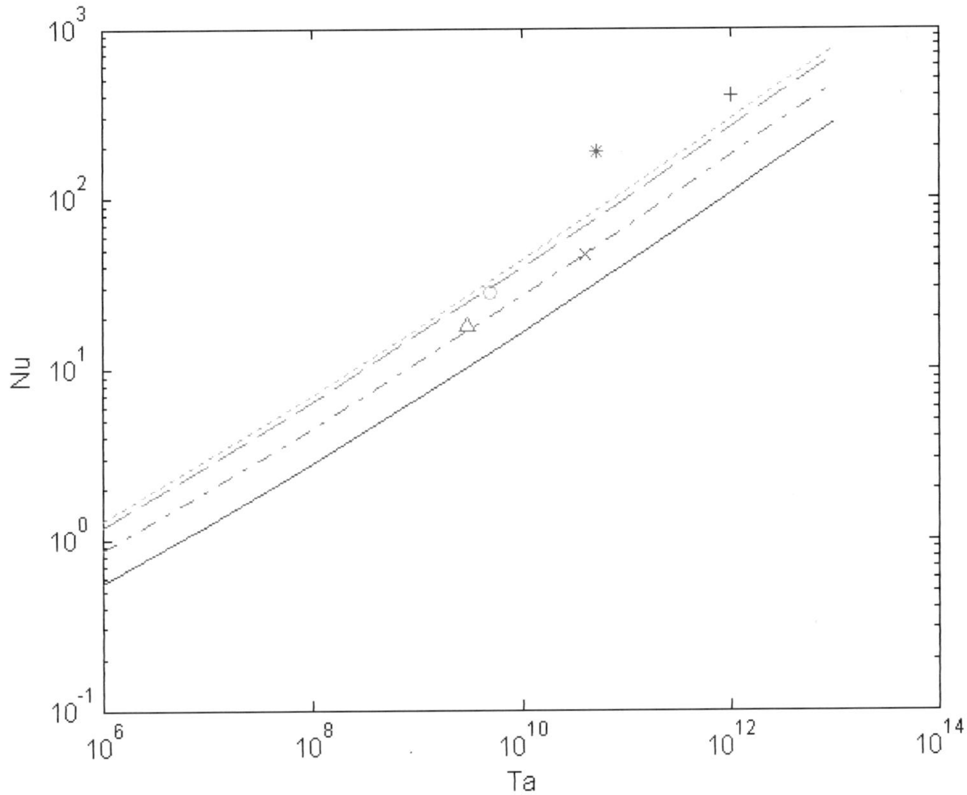

Figure 1. Dependence Nusselt number (Nu) versus Taylor number (Ta) for $\eta=0.357$ - triangle (Δ) [11]; for $\eta=0.5$ - cross (x) [12], circle (O) [13]; for $\eta=0.716$ - plus sign (+) [14]; for $\eta=0.909$ - star (*) [15]. Theoretical curves for $\eta=0.357; 0.5; 0.716; 0.909$ are from bottom to top.

We see that theoretical dependences of Nusselt numbers on Taylor numbers partly coincides with results [2,3], but our curves are lower than experimental data [11-15]. One can be assumed drawbacks of our approach, but there is a good agreement of theoretical values of torque with the experiments [16], [17]. Comparisons was made in our papers [7], [5] correspondingly.

Dependence Nusselt number on radius is depicted on Figure 2.

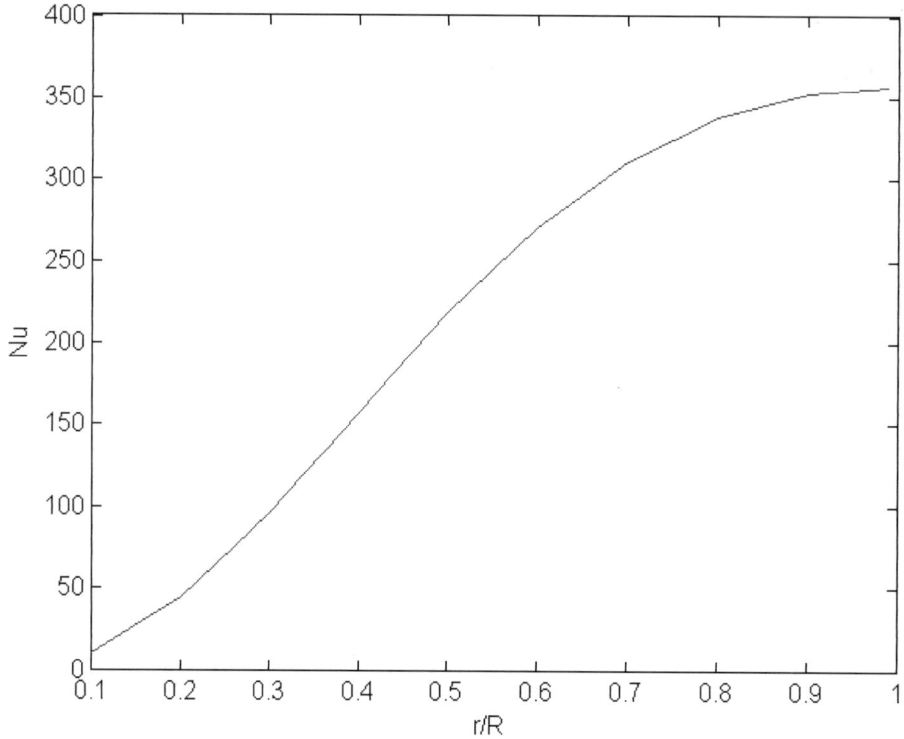

Figure 2. Dependence Nusselt number on radius ratio $r/R=\eta$, $Ta=10^{11}$.

There is some agreement with the models [2].

4. Conclusion

It seems that we should look for some fundamental theory of turbulence. The theory must explain spectra of turbulence from large scales to dissipative ones. Then we can create the turbulence models as a system of differential equations. A success of an approach should be verified for Taylor- Couette flow and Poiseuille flow. After that we can deal with more complex turbulent flows.

References

[1] Zhou Ye 2021 *Physics Reports* **935** 1-117
[2] Cheng W, Pullin D I and Samtaney R 2020 *J. Fluid Mech.* 890 A17
[3] Berghout P, Verzicco R, Stevens R J A M , Lohse D, Chung D 2020 *J. Fluid Mech.* **905** A11
[4] Balonishnikov A M 2000 *Phys. Rev.* E **61** **2** 1390-4
[5] Balonishnikov A M 2016 *Tech Phys* **61** **11** 1755-8
[6] Balonishnikov A M, Kruchkova J V and Opalikhina O V 2020 *IOP Conf. Ser.: Mater. Sci. Eng.* **862** 022029
[7] Balonishnikov A M and Kruchkova J V 2020 *IOP Conf. Ser.:* **1679** 022062
[8] Balonishnikov A M, Korshakova I A and Kruchkova 2021 *IOP Conf. Ser.:* **1889** 022105
[9] Dubrulle B and Hersant F *2002 Eur. Phys. J.* B **26** 379
[10] van Gils D P M, Huisman S G, Grossman S, Sun C and Lohse D 2012 *J. Fluid. Mech.* **706** 118-49
[11] Froitzhheim A, Merbold S R, Ostilla-Monico R, and Egbers C 2019 *Phys. Rev. Fluids* **4** 084605
[12] Ostilla-Monico R , van der Poel E P , Verzicco R, Grossman S and Lohse D 2014 *J. Fluid. Mech.*

761 1-26
[13] van der Veen R C, Huisman S G, Merbold S, Harlandre W, Egbers C, Lohse D and Sun C 2016 *J. Fluid. Mech.* **799** 334-51
[14] van Gils D P M , Huisman S G , Bruggert G W, Sun C and Lohse D 2011 *Phys. Rev. Lett.* **106** 024502
[15] Ostilla-Monico, R, Verzicco R., Grossman S., Lohse D 2015 *J. Fluid Mech.* **788** 95-117
[16] Lewis G S and Swinney H L 1999 *Phys. Rev.* E **59** **5** 5457-68
[17] Paoletti M S and Lathrop D P 2011 *Phys. Rev. Lett.* **106** 024501

Development of design solutions for improving the characteristics of capacitive sensitive elements of sensors of physical quantities

P G Mikhailov

Penza State Technological University, Penza, Russian Federation

E-mail: pit_mix@mail.ru

Abstract. Capacitive measuring transducers are the most stable of the sensors of physical quantities, so they are used in such areas and on products in which the parameters of time and parametric stability are the main parameters. Such products and systems include rocket and space technology, aviation, military equipment and weapons. In addition, capacitive measuring transducers are energy-saving, since the sensor element does not consume electrical energy during operation. To achieve high time and parametric stability, the design of measuring transducers must be designed using certain design and technological solutions aimed at eliminating the occurrence of unintentional internal mechanical stresses, and compensating for thermal stresses. In addition, capacitive measuring transducers must be guaranteed to be sealed, and also during their manufacture, all parts undergo a degassing operation, in which moisture dissolved in the near-surface layers of materials is removed by heat treatment in vacuum, as well as using methods of heterization of vacuum cavities.

1. Introduction

The fundamental conditions for creating technically advanced capacitive sensors (CS) are:

$$\frac{W}{\ell_1 + \ell_2} \to 1 \quad \text{and} \quad \frac{W}{h_0} \to 1$$

where is the deflection of the membrane, l1, l2 and h0 are the geometric parameters of the membranes.

From the analysis of the above conditions, it follows that it is necessary to strive to increase both the absolute and relative deflection of the membrane elastic element of the pressure sensor.

The deviation of the rigid center of the membrane is equal to [1]

$$W = A_p \frac{PR^4}{64D}, \tag{1}$$

where $D = \frac{E t_i^3}{12(1-\mu^2)}$ is the stiffness of the membrane; $A_p = (1 - \gamma^4 + 4\gamma^2 \ln \gamma)$ - normalizing factor, r - the radius of the rigid center of the membrane.

IOP Conf. Series: Materials Science and Engineering 1227 (2022) 012002 doi:10.1088/1757-899X/1227/1/012002

The W deflection W must satisfy the condition $\sigma_{max} \leq [\sigma]$, where σ_{max} - maximum stress in the membrane at the surface points on the outer and inner contours of the elastic part; $[\sigma]$- the permissible stress for the membrane material.

$$\sigma_{max} = B_p \frac{E t_i W}{R^2}, \tag{2}$$

where $B_p = \frac{4}{1-\mu^2} \cdot \frac{1-h^2}{1-\gamma^4+4\gamma^2 \ln \gamma}$.

The maximum possible deviation of the membrane is [2]:

$$W_{max} = A_p \frac{R[\sigma]^{3/2}(1-\mu^2)}{P^{1/2}} \tag{3}$$

where $A_p = \frac{1-\gamma^4+4\gamma^2 \ln \gamma}{2\sqrt{3}(1-\gamma)^{3/2}}$.

From the analysis of this relationship, it follows that an increase in the absolute value of the deflection can be achieved by increasing the size of the membrane and choosing a membrane material that corresponds to the complex indicator

$$A_p = \frac{1-\gamma^4+4\gamma^2 \ln \gamma}{2\sqrt{3}(1-\gamma)^{3/2}}, \tag{4}$$

which combines the mechanical characteristics of the selected material.

2. Materials and methods

By selecting a material with a maximum value of ψ, you can increase the deflection by 3-4 times. However, this conclusion is valid only if the membrane material is not subject to important requirements, such as: high corrosion resistance, vacuum density, weldability of various types of welds, polishability up to 13-14 kl. Purity, low temperature coefficient of linear expansion (TCLE) and, most importantly, low temperature coefficient of elastic modulus β (TKEM) in a wide temperature range, since it is the value of this coefficient that determines the temperature multiplicative error of the developed capacitive quasi-differential pressure sensors [3, 4].

Taking into account the latter condition, a complex indicator that determines not only the deflection of the membrane, but also its heat resistance, can be presented in the form:

$$\psi' = \frac{[\sigma]^{3/2}(1-\mu^2)}{E\beta}. \tag{5}$$

Then, of course, when choosing a structural material for elastic elements of capacitive pressure sensors, preference should be given to elinvar alloys, such as типа 29N26KHTBYU, EP-920, VUS-22, which have the maximum values of the indicator ψ': $4 \cdot 10^6$, $2{,}5 \cdot 10^6$, $22 \cdot 10^6$ [5]. In this case, it is possible to reduce the temperature error of pressure sensors by an order of magnitude in comparison with sensors made of 36 NHTYU alloy ($\psi' = 0.5 \cdot 10^6$). An increase in membrane deflection can be achieved due to the planned radial profiling of the membrane thickness (figure 1) according to the expression [6]

$$\frac{t_i}{t'_i} = e^{\frac{\varphi}{\sigma}\left(\frac{r}{R}\right)^2}, \tag{6}$$

where φ is a constant value; σ is the elastic limit.

The deflection in the center of such a membrane (figure 1a) can be represented by the formula:

$$W_0' = \eta \frac{6(1-\mu^2)R^4 P}{E t_i'}, \tag{7}$$

where $\eta = f(\varphi)$ - the value of this function at μ=0.3 for negative values of φ is given in table 1.

The graph of the dependence of this ratio on the coefficient ψ is shown in figure 1b: $f(\varphi) = W_0' / W_0''$

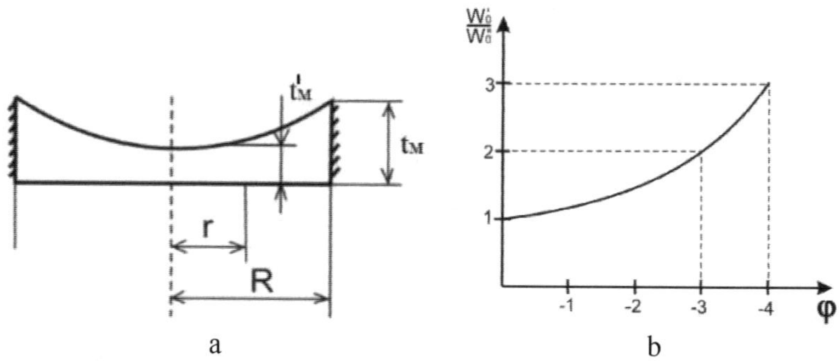

Figure 1. Membrane of variable thickness (a) and its load characteristic (b).

Table 1. Function values for μ=0.3.

φ	0	-1	-2	-3	-4	-5
ή	0,0313	0,0246	0,0192	0,0152	0,0152	0,0119

Let us compare the deflections of a profiled membrane calculated from (7) and a flat membrane with a thickness t_i (thickness in the seal).

Deflection of the flat membrane:

$$W_0'' = \frac{3PR^2(1-\mu^2)}{16Et_i^3} = \frac{3PR^2(1-\mu^2)}{16Et_i^3 e^{-\varphi/\sigma}}. \tag{8}$$

It follows from the graph that by profiling the membrane in thickness, it is possible to increase the deflection by almost 3 times.

Similar studies of a membrane with a rigid center have shown that the profiling effect is reduced several times, but remains significant (30%) at the ratio r/R=0,2.

In the practice of creating sensors, there are requirements for maintaining metrological characteristics (MC) after exposure to overload pressures from 150 to 300% of the nominal ones. This requirement in sensors is most often achieved by calculating the membrane with a 50% margin for deviation and initial clearance.

The rejection of reserves for deflection and initial clearance (while ensuring the sensor's operability in the event of overload in another way) reduces the measurement error by about 1.5 times.

Summing up the results of the conducted research, it can be concluded that by selecting a structural material with the maximum value of the complex index ψ', as well as by profiling the thickness of the elastic element of the capacitive pressure sensor and reducing the deflection margin, it is potentially possible to reduce the measurement error many times (by 30-40 times) [7-9].

3. Analytical estimation of the temperature error of a quasi-differential thin-film capacitive pressure sensor

The process of converting pressure P to the ratio of capacities is carried C_0/C_x out in two stages:

1. Conversion of pressure to deflection (displacement) of the membrane (working electrode) P(W).
2. Conversion of displacement to the ratio of capacities W (C_0/C_x).

A change in temperature, as noted earlier, can lead to the following changes in the capacitive sensing element (CSE) parameters [10]:

- pressure in the zamembrane cavity,
- geometric dimensions of the membrane,
- elastic modulus of the membrane material,
- the dielectric constant of the medium between the electrodes,
- the area of the electrodes,
- the gap between the electrodes.

Changes in these parameters, in turn, cause the corresponding temperature conversion errors.

We will analyze the temperature errors for the above two stages of transformation using the example of an CSE with a membrane having a rigid center.

4. Analysis of the error caused by the change in pressure in the zamembrane cavity

In our opinion, it is advisable to analyze three possible physical states in practice of the transmembrane cavity CSE:

1. The zamembrane cavity is not sealed and interacts with the external environment, which can be very different in its properties, chemically neutral or aggressive.
2. The membrane cavity is sealed from the external environment, but contains dry air under pressure P_0. Since the possible temperature change lies in the range from minus 253 to + 640 °C, almost all the main gas components located in the zamembrane cavity at a temperature of minus 253 °C will pass from the gas to the liquid and solid phases. In this case, there will be a sharp increase in their dielectric constant (up to about 1,4).

Based on physical considerations, it is not difficult to see that the most likely place of gas condensation during the pressure conversion of cryogenic components is the electrode located on the membrane. It is for this reason that successive sharp changes in the output signal can be observed. Hence, it is natural to conclude that both of the considered states of the CSE zamembrane cavity are practically unacceptable for solving the problem. In order to avoid the occurrence of the considered temperature error, it is necessary to perform deep vacuum evacuation of the CSE membrane cavity to the value of the residual pressure significantly lower than the converted one.

In practice, this requirement for a number of technical reasons may not be completely feasible, then the value of the residual pressure should be determined based on the permissible value of the temperature error.

The relationship between pressure, temperature, and volume is described by the Mendeleev-Klayperon equation, one of the most simplified forms of writing, which has the form:

$$PV = P_0V_0\frac{T}{T_0} \tag{9}$$

where is P_0 the pressure at temperature T_0=273,1 K; V_0 -is the volume occupied by a gram-mole of gas at T_0; T - absolute temperature; P - gas pressure; V - gas volume.

If we neglect the change in volume V and pressure P in the zamembrane cavity due to the movement of the membrane, then according to when the temperature changes by an amoun $\Delta T = T - T_0$, a corresponding change in pressure

$\Delta P = P_0\frac{\Delta T}{T_0}$, then it will be determined by the relation $P = P_0 + P_0\frac{\Delta T}{T_0} = P_0\left(1+\frac{\Delta T}{T_0}\right)$

Dividing both parts of the expression by the value P_x (nominal pressure), we get

$$\frac{\Delta P}{P_x} = \frac{P_0}{P_x}\cdot\frac{\Delta T}{T_0} \tag{10}$$

Given that $\Delta P/P_x$ there is nothing but a relative reduced error, we find

$$\gamma_p = \frac{P_0}{P_x} \cdot \frac{\Delta T}{T_0} \tag{11}$$

The error γ_p caused by the change in pressure in the zamembrane cavity is additive, and its relative influence increases with a decrease in the pressure range. Based on this analysis, we conclude that at low pressures, the zamembrane cavity must be evacuated.

5. Analysis of the error caused by changes in the elastic modulus of the membrane material and its geometric dimensions

The error under consideration occurs at the stage of converting the pressure into the movement of the electrode. $P_x(W)$. The expression for the displacement of a membrane with a rigid center can also be represented as [11]:

$$W_0 = A_p \frac{R_0^4}{E_0 t_0^3} P_x , \tag{12}$$

where $A_p = \frac{3 \cdot (1-\mu^2) \cdot \gamma_0^4 - 1 - 4\gamma_0^2 \ln \gamma_0}{16\gamma_0^4}$; $\gamma_0 = \frac{r_0}{R_0}$; R_0 - the outer radius of the membrane; r_0 - the radius of the rigid center; t_{i0} - the thickness of the membrane outside the rigid center.

Let us $A_p \cdot \frac{R_0^4}{E_0 t_0^3} = A'_p$ denote, then the expression (12) can be represented as

$$W_0 = A'_p \cdot P_x \tag{13}$$

When the temperature changes by ΔT, the parameters E, R, r, t_i in the framework of the linear theory of thermal conductivity will be determined by the expressions $E = E_0(1+\beta\Delta T)$, $R = R_0(1+\alpha\Delta T)$, $r = r_0(1+\alpha\Delta T)$, $t_i = t_{i\,0}(1+\alpha\Delta T)$, where α is the TCLE of the membrane material; β is TKEM of the membrane material.

The expression for the deflection of the membrane, taking into account the pressure in the zamembrane cavity, but without taking into account the temperature change, will have the form

$$W_{0P_0} = A'_p(P_x - P_0) \tag{14}$$

Based on (12), taking into account (13) changes in temperature and pressure in the membrane cavity, the expression for the membrane deflection can be written

$$W = A'_p \frac{1+\alpha\Delta T}{1+\beta\Delta T}\left[P_x - P_0\left(1+\frac{\Delta T}{T_0}\right)\right] \tag{15}$$

or

$$W = \frac{1+\alpha\Delta T}{1+\beta\Delta T}\left(W_0 - W_{P_0} - \Delta W_{P_0}\right), \tag{16}$$

where $W_{P_0} = A'_p \cdot P_0$ is the deflection of the membrane under the influence of the residual pressure of the zamembrane cavity; $\Delta W_{P_0} = A'_p \cdot P_0 \frac{\Delta T}{T_0}$ - change the deviation W_{P0} from the temperature change to ΔT.

The relative change in the deflection taking into account the change in temperature and pressure in the zamembrane cavity will be equal

$$\frac{\Delta W}{W_{0po}} = \frac{W - W_{0po}}{W_{0po}} = \frac{A'_p \frac{1+\alpha\Delta T}{1+\beta\Delta T}\left[P_x - P_0\left(1+\frac{\Delta T}{T_0}\right)\right] - A'_p(P_x - P_0)}{A'_p(P_x - P_0)} = \frac{\frac{1+\alpha\Delta T}{1+\beta\Delta T}\left[P_x - P_0\left(1+\frac{\Delta T}{T_0}\right)\right]}{P_x - P_0} \tag{17}$$

6. Results and conclusion

Analyzing the expression (17), it is easy to see that when the temperature changes, some compensation for the change in the deflection of the membrane occurs, due to changes in the parameters α and β due to changes in the pressure in the cavity behind the membrane. When choosing a structural material for the membrane, the determining factor is the parameter β. It is important not only to select a low β construction material, but also to normalize it over the entire operating temperature range [12].

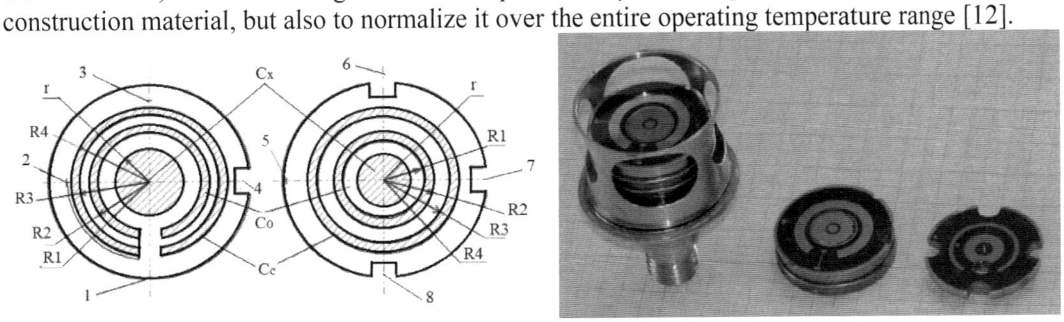

a b

Figure 2. Topology (a) and photography (b) of differential CSE: 1 и 3; 6 и 8; 5 и 7; 2 и 4 – topological axes CSE; R1-R5 - film electrode radii; C_x, C_0, C_c - capacitive elements, r – радиус of the CSE.

To ensure the parametric and temporal stability of capacitive sensors, mathematical models and design solutions for CSE are proposed, which ensure the physical and design compatibility of materials and assemblies (figure 2).

References

[1] Jackson R G 2007 *The latest sensors from English* (M: Technosphere) 380 p
[2] Vavilov V D and Timoshenkov S P 2018 *Microsystem sensors of physical quantities monograph in 2 parts* (Moscow: Technosphere) 550 p
[3] Zhang Yu, Hauwer R and Yazdi N 2011. Highly sensitive ultra-thin capacitive MEMS pressure sensor *IEEE International Conference Journal of Engineering Science and Technology March 2016,* **11(3)** pp 112-5
[4] Mikhailov P G, Chuvykin B V and Mikhailov A P 2020 Modeling of dimensional schemes of pressure control devices. *Physics journal: Conference series* **1515** 042099 doi: 10.1088 / 1742-6596 / 1515/4/042099
[5] Abdelaziz Heddia, Fuad Kerrur and Salah Kemush 2016 Thermal Drift Characteristics of Capacitive Pressure Sensors *Journal of Engineering Science and Technology* **11(3)** 346-61
[6] Shahiri M and Tabarestani B 2012. Development and modeling of a highly sensitive capacitive pressure sensor with a slot diaphragm. *International Conference on Biomedical Engineering* pp 484-9
[7] Nurul Amzia, MD Yunus, Ijal Abdul Khalin and Nasri Suleiman 2015. Evaluation of MEMS Pressure Sensors and Device Applications World Academy of Sciences, *Engineering and Technology International Journal of Electronics and Communication Technology* **9(8)**
[8] Fernandez J M, Bonet N, Siiro J J and Lopez J M 2013. Ceramic Capacitive Pressure Transmitter Based on LTCC Technology, at *IEEE Spanish Conference pp* 111-4
[9] Grant T, Joshi W and Sviystrup H 2014 Measuring the variability of time from sitting to standing using pressure transducer technology mattress at the *IEEE International Symposium.*
[10] Chaurazia S 2013 Analytical Model of the NEMS Square Diaphragm Piezoresistive Pressure Transmitter *IEEE Engineering and Systems Student Conference.*
[11] Zhang Yu, Howver R and Gogoi B 2011. Journal of Engineering Science and Technology March 2016 **11(3)** MEMS High Sensitivity Ultra-Thin Capacitive Pressure Sensor *IEEE*

International Conference on Solid State Sensors pp 112-5
[12] SlesarevYu N, and Vorontsov A A 2021 Analysis of models of solenoids used to simulate the reproduction signal of magnetostrictive tiltmeters. *Proceedings - nternational Conference on Industrial Engineering, Applications and Manufacturing, ICIEAM 2021 pp* 897-902

Quasioptimal control in diversed signal transmission

P A Polushin, O R Nikitin[1] and I R Dubov

Vladimir state University, Vladimir, 600000, Russia

[1]E-mail: olnikitin@mail.ru

Abstract. To increase the noise immunity of signal transmission, diversity methods are now widely used, consisting in obtaining and combining several copies of the transmitted signal. In this case, it is possible to perform a combination either before the detection procedure or after it. If you do not take into account the possible use of non-linear types of modulation, then the pre-detector combination always has advantages over the post-detector combination. However, taking into account the nonlinear properties of the transmitted signals, new possibilities appear for increasing the noise immunity in combination and simplifying the processing. In the case of using analog signals, in particular frequency modulation, at certain points in time, the pre-detection combination can lose to the post-detection combination. At the same time, by combining pre-detector and post-detector combining circuits, it is possible to lower the threshold level during demodulation and increase noise immunity. In the case of using digital modes of modulation, it is possible to process only the signals after demodulation without reducing the noise immunity and to eliminate the need for preliminary phasing of the diversity signals before detection.

1. Formulation of the problem

Diversity methods have been known for a long time, but even now they are quite widely used in signal transmission ([1-11]) Space diversity, despite the fact that it requires the use of several antennas, can significantly improve the noise immunity of signal transmission when transmitting over non-stationary channels. With the correct organization of the communication system, fading in various channels occurs independently, which makes it possible to reduce the likelihood of communication disruptions.

Diversity techniques are based on the transmission and reception of multiple identical copies of a signal and their sharing. The received copies must be combined in order to obtain the maximum gain in noise immunity. In this case, a number of peculiarities arise, in particular, the combination can be performed both before and after detection. However, when using the features of the structure of the transmitted signals, it becomes possible to obtain an additional gain in transmission noise immunity.

It is a generally accepted fact that the pre-detection combination is superior to the post-detection one ([2-4]). Provided that the useful components are preliminarily phased, they add up "in amplitude", while noises with random mutual phase shifts add up "in power". This increases the signal-to-noise ratio before demodulation. However, the non-linear properties of the transmitted signals can make their own "corrections". The same is true for both analog and digital types of modulation.

2. Implementation of nonlinear features of the signals used

Among the nonlinear types of modulation, features can be observed that indicate the temporal advantages of post-detector addition over pre-detector addition ([3-5]). Let be γ_{in}- instantaneous ratio of signal voltage and noise before detection; γ_{out} - the same ratio after detection. Let us denote through $L\{\bullet\}$ relationship between γ_{in} and γ_{out} for this type of modulation, $\gamma_{out} = L\{\gamma_{in}\}$. We also denote $Q\{\bullet\}$ by denote the operation of determining the signal-to-noise ratio after combining the diversity signals, in which the signal-to-noise ratio values are equal $\gamma_1 \div \gamma_N$, i.e. $\gamma_{out} = Q\{\gamma_1, \dots, \gamma_N\}$. Thus, as a result of pre-detector addition, the resulting signal-to-noise ratio will be equal, and as a result of post-detector addition $\gamma_D = L\{Q\{\gamma_1, \dots, \gamma_N\}\}$, the result will be $\gamma_P = Q\{L\{\gamma_1\}, L\{\gamma_2\}, \dots, L\{\gamma_N\}\}$. It can be shown that instantaneous values γ_D are not always greater than instantaneous values γ_P.

As an example, when using analog signals, consider frequency modulation. The result of summing the received signal $U(t)$ and $n(t)$ noise can be represented in the form $x(t) = U(t) + n_c(t) + n_s(t)$, where $n_c(t)$-noise component in phase with the useful signal, $n_s(t)$-noise component orthogonal to the useful signal. At FM, the value of the amplitude U_c of the useful signal during an interval of time less than the quasi-period of fading can be considered constant ([2-6]). Orthogonal noise components; $n_c = n_a\, cos\,\varphi_n$; $n_s = n_a\, sin\,\varphi_n$ mutually independent. The instantaneous phase changes φ_n are the sum of the independent instantaneous phase changes of the original noise and the phase change of the useful signal, i.e. distribution φ_n can be considered uniform within $0 \div 2\pi$. Thus, the error of the instantaneous phase of the sum of the useful signal and noise with respect to the phase of the useful signal will be equal to

$$\varphi_1 = arctg\frac{n_a \sin\gamma_n}{U_c + n_a \cos\gamma_n} = arctg\frac{a\sin\gamma_n}{1 + a\cos\gamma_n},$$

where $a = \frac{n_a}{U_c}$.

As a result of frequency detection, this instantaneous phase error will lead to an instantaneous frequency error that determines the noise after detection, which is determined by the formula:

$$\varepsilon_1 = \frac{d\varphi_1}{dt} = a\frac{a + \cos\varphi_n}{1 + 2a\cos\varphi_n + a^2}\frac{d\varphi_n}{dt}$$

The quantities φ_n and $\frac{d\varphi_n}{dt}$ are mutually independent, just like the quantities a and φ_n. Let's single out the coefficient

$$b_1 = a\frac{a + \cos\varphi_n}{1 + 2a\cos\varphi_n + a^2}$$

This factor will determine the degree of transformation of the instantaneous frequency of the noise before detection to the instantaneous frequency of the noise after the detection. Let's average it over the phase taking into account its uniform distribution in the interval $0 \div 2\pi$. We will also perform averaging in the root-mean-square sense because of the alternation of signs b_1.

$$\beta_1 = a\sqrt{\frac{1}{2\pi}\int_0^{2\pi}\left(\frac{a + \cos\varphi_n}{1 + 2a\cos\varphi_n + a^2}\right)^2 d\varphi_n}$$

As a result of mathematical transformations, we obtain

$$\beta_1 = \begin{cases} \dfrac{a}{\sqrt{2}\sqrt{1-a^2}}; |a| < 1 \\ \sqrt{\dfrac{2a^2-1}{2(a^2-1)}}; |a| > 1 \end{cases}$$

The obtained expressions correspond to the situation of ideal frequency detection, when the amplitude fluctuations at the BH input are completely removed from its result, i.e. there is a deep signal limitation.

After calculating the RMS value by phase averaging, i.e.

$$\beta_2 = \sqrt{\frac{1}{2\pi}\int_0^{2\pi} b_2^{\,2} d\varphi_n}$$

as a result we have

$$\beta_2 = \begin{cases} \dfrac{a}{\sqrt{2}}; |a| < 1 \\ \sqrt{a^2 - 0{,}5}; |a| > 1 \end{cases}$$

The clipping method (which is usually always done before frequency detection) and the clipping level can be different. Therefore, the dependence of the corresponding averaging result for a real circuit will lie between the graphs $\beta_1(a)$ and $\beta_2(a)$. Similar graphs are shown in figure 1. The graphs show the dependences: 1 - for $\beta_1(a)$, 2 – for $\beta_2(a)$. When the signal level falls below the limiting threshold, it is necessary to switch from the shape of the graph 1 to the shape of the graph 2 with a certain coupling coefficient before $\beta_2(a)$, so that during this transition there are no gaps in the instantaneous values of the signals.

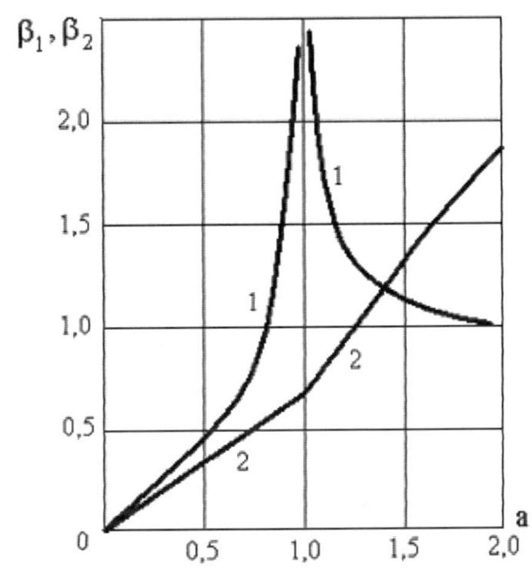

Figure 1. Threshold function.

Figure 1 shows that the dependence $\gamma_{out} = L\{\gamma_{in}\}$ at FM is by no means monotonic. It has an area near the value $a = 1$ where the noise level rises sharply. As is known [3, 6], this manifests itself in the appearance of the so-called "anomalous" noise in the form of impulse surges. As the input signal-to-noise ratio deteriorates, the emission rate increases rapidly, so the output signal-to-noise ratio falls faster than the input signal-to-noise ratio. This manifests itself as a known threshold in FM demodulation. This phenomenon is usually explained as follows (figures 2, 3 and 4). The explanation is that for close values of the vectors of the signal U and noise n, their total vector y = U + n rotates around the center of coordinates very quickly, even if the rate of rotation of the vector n is small. This takes place when the end of the vector y falls into some region Ω near the origin (for n ≈ –U).

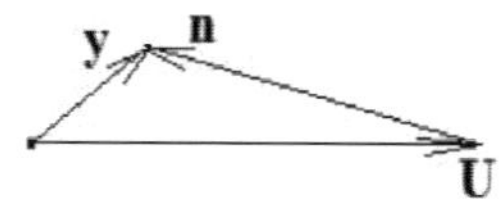

Figure 2. Vector sum of signal and noise.

Such a rapid turn turns into an overshoot as a result of frequency detection. When the level of the useful signal decreases (for example, as a result of fading), the probability that the condition | n | ≈ | U | increases, so the frequency of occurrence of emissions also increases.

In the case of diversity reception, the absence of monotonicity of the function $L\{\bullet\}$ provides additional opportunities for improving noise immunity as a result of lowering the threshold level. This is due to the fact that in this case the condition is $L\{Q\{\gamma_1,...,\gamma_N\}\} > Q\{L\{\gamma_1\},...,L\{\gamma_N\}\}$ not always met. This is illustrated in figure 3 for a diversity factor of N = 2. (The addition is assumed to be linear.)

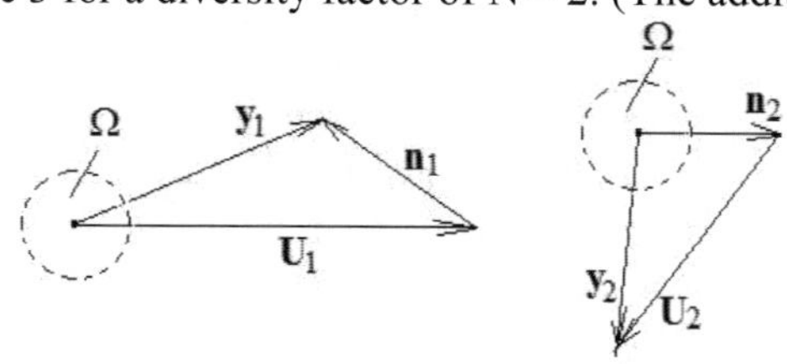

Figure 3. Vector sum for diversity factor N=2.

Figure 3 shows vector diagrams of signals in two diversity branches. The total vectors y1 = U1 + n1 and y2 = U2 + n2 in both cases separately do not fall into the region Ω corresponding to the appearance of an outlier, i.e. after frequency detection, no spike occurs in any of the branches. Figure 4 shows vector diagrams of signals after pre-detector addition. In this case, the total vector U1 + U2 + n1 + n2 falls into the region Ω, as a result of which, after detection, an overshoot will take place.

Thus, with such a combination of signal and noise vectors, the pre-detector addition of signals generates an overshoot, while the post-detector addition does not. (Although, in general, the total number of emissions during the post-vector addition is naturally greater than during the pre-detector addition, however, their appearance basically does not coincide in both types).

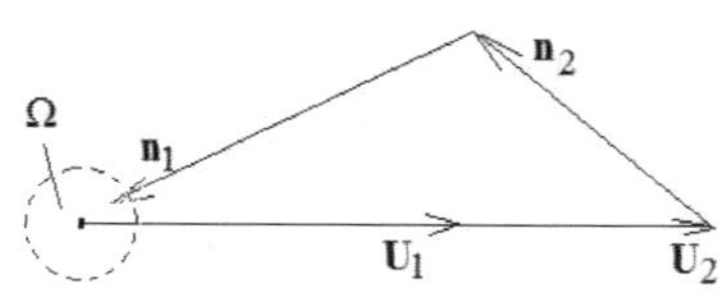

Figure 4. Situation of overshoot.

The considered phenomenon can be used to eliminate part of the surges during pre-detector addition, if at the same time the post-detector addition signal is used for this time. To do this, consider the circuit shown in figure 5.

At the moments of time corresponding to the presence of an anomalous noise emission in the pre-detector addition signal and its absence in the post-detector addition signal, the post-detector addition signal is switched to the output, and the rest of the time - to the pre-detector addition signal. In the adder (S2), the pre-detector diversity combining is carried out (it is assumed that they are phased before). In frequency detectors, each diversity signal is detected separately (FD1 and FD2) and their sum (FD3). In the adder (S1), post-detector addition is performed. All frequency detectors are assumed to be sufficiently identical, i.e. in the absence of noise, their output useful signals are the same.

In subtraction blocks (-), the signals of each branch are subtracted from the total signal. The received voltages are fed to the input of threshold devices (PU1 and PU2), which generate a logic signal "1" if there is an overshoot at the input, and a logic signal "0" if there is no overshoot. These threshold devices determine the voltage level that can be considered an overshoot. The logical signals of the threshold devices are combined by the logical operation "AND", the result of which controls the operation of the switch (K). Delay elements (LZ) are used to compensate for the time delay in switching the switch due to the finite response time of the threshold devices and ensure the simultaneous arrival of surges and the control signal to the switch.

Let us denote the probabilities of the appearance of outliers in the first channel, in the second channel, and in the case of pre-detector addition through, respectively $P\{y_1 \in \Omega\}$, $P\{y_2 \in \Omega\}$ and $P\{y_1 + y_2 \in \Omega\}$, as the probabilities of getting the ends of the vectors, y_1, $\mathbf{y}_2$ and $y_1 + \mathbf{y}_2$ into the region Ω. Hence, we obtain the probability of an outlier occurring during post-detector addition:

$$P_{AD} = P\{y_1 \in \Omega\} + P\{y_2 \in \Omega\} - P\{y_1 \in \Omega\}P\{y_2 \in \Omega\}$$

The probability of an outlier in the circuit in figure 5.

$$P_C = P\{\mathbf{y}_1 \in \Omega\}P\{\mathbf{y}_2 \in \Omega\}$$

Disregarding emission levels, inequalities can guide the relative effectiveness of schemes:

$$P\{\mathbf{y}_1 \in \Omega\} + P\{\mathbf{y}_2 \in \Omega\} - P(\mathbf{y}_1 \in \Omega\}P(\mathbf{y}_2 \in \Omega\} >\to$$
$$\to> P\{\mathbf{y}_1 + \mathbf{y}_2 \in \Omega\} >> P\{\mathbf{y}_1 \in \Omega\}P\{\mathbf{y}_2 \in \Omega\}$$

Thus, the circuit in figure 5 allows to eliminate a large number of surges at the FD output during pre-detector addition, thereby significantly reducing the detection threshold level. The magnitude of the threshold level reduction is determined by many factors, such as the level of limitation before detection, the frequency modulation index, the shape of the signal spectrum, etc. Computer simulation has shown the possibility of lowering the threshold level in different conditions by 3.7 dB. Efficiency in practice will be determined largely also by technical characteristics, mainly by the speed of the switch at the output of the circuit.

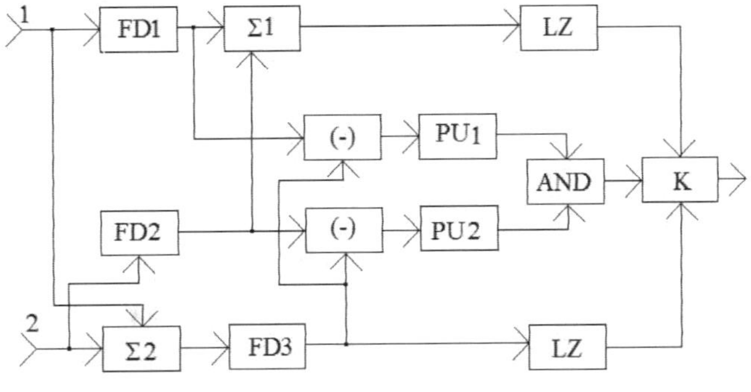

Figure 5. Ciruit for diversity factor N=2.

A similar approach can be applied for the diversity factor N> 2. In this case, in a simple version, the switch connects to the output either the signal obtained after the FD general pre-detector addition, or the post-detector addition signal of all separately detected diversity signals. In a more sophisticated version, a multi-input switch can be used. It selects the best signal from several combinations of input signals obtained by summing with different sets of weights.

With the use of digital signals, diversity as a technique has retained its relevance. At the same time, it became possible to implement the combination much easier. To clarify this, it is necessary to consider the transmission of BPSK signals or any other binary signals that, upon detection, take two logical values. It is assumed that the analog value of the signal is compared with a certain level, as a result of which one of two possible logical values is selected.

When detection is performed by two separate detectors, their output signals may or may not coincide. If the output signals of the detectors match, it means that either there is no error in both signals, or both signals are affected by an error. Naturally, the overall result of the analysis will also be erroneous.

Let's analyze 2-fold diversity. Let $s_1 = u_1 + n_1$ and $s_2 = u_2 + n_2$ denote the signals arriving at the detectors of the first and second branches, where: u_1, u_2 are useful components, n_1, n_2 are the noise components of the first and second diversity signals. When using addition before the detector, the signal in front of the detector will be $s_C = (u_1 + u_2) + (n_1 + n_2) = u_C + n_C$. In practice, the power of both binary signals is the same, and the appearance of both variants is equally probable. Without taking into account noise, signal "1" corresponds to level «a» at the detector input, signal "0" corresponds to level «$-a$». The decision is made by comparing the obtained level with zero. Let the character "0" be transmitted. (If a symbol equal to a logical one is transmitted, all reasoning is similar). An error occurs on the first diversity channel if $n_1 > a$. In the second diversity channel, it will occur if $n_2 > a$. At the same time, an error during addition to the detector will occur if $n_1 + n_2 = n_C > 2a$.

To implement the described quasi-optimal control, the scheme shown in figure 1 is proposed.

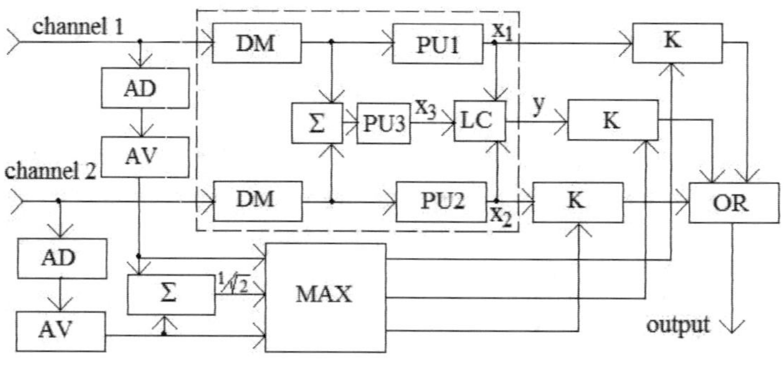

Figure 6. Ciruit for diversity factor N>2.

The section of the diagram, highlighted by shading, works as follows. During post-detector addition, the diversity signals in the first and second diversity channels are processed in demodulators (DM), as a result, voltages s_1 and s_2 are generated, which are fed to the first and second threshold devices (PU1 and PU2). In these devices, the obtained voltages are compared with the zero level, and based on the comparison results, logical signals x_1 and x_2 are generated. Signal "1" is generated if the signals are above zero, signal "0" - when they are below zero. Simultaneously, the adder (Σ) determines their sum $s_3 = s_1 + s_2$. The resulting sum in the third threshold device (PU3) is also compared to zero. Based on this, according to the same rule, the signal $x_3 = 1$ is generated if the sum is greater than zero, and the signal $x_3 = 0$ if the sum is less than zero. When x_1 and x_2 are different, the signal x_3 "tilts towards" that of them, whose voltage at the output of the demodulator is greater in absolute value. The received three logical signals go to the logic circuit (LC). In it, on their basis, a logical function y of the following form is developed:

$$y = (x_1 x_2 \vee \bar{x}_1 \bar{x}_2) x_1 \vee \overline{(x_1 x_2 \vee \bar{x}_1 \bar{x}_2)} x_3 = x_1 x_2 \vee x_1 \bar{x}_2 x_3 \vee \bar{x}_1 x_2 x_3.$$

When signals x_1 and x_2 are the same, any of these signals (let signal x_1) go to the output. When signals x_1 and x_2 are different, signal x_3 is output. As a result, the following result is observed. Let a signal equal to -a be transmitted at some moment in time, which corresponds to the transmission of a logical zero. An error can be when, or both PU generate the same signal "1", or both PU generate different signals, but the wrong one will be selected.

Let's analyze the last situation in more detail. Let the signal values be equal: $x_1 = 1$ and $x_2 = 0$, (or $x_1 = 0$ and $x_2 = 1$). In this case, the decision will be made based on the value of the variable x_3. An error will occur if a signal $x_3 = 1$ is generated and a decision is made that a logical unit was transmitted. This situation corresponds to the fulfillment of the inequality: $s_1 + s_2 > 0$, which means that either there are inequalities $n_1 - a + n_2 - a > 0$, or $n_1 + n_2 > 2a$. In this case, the "boundary" between correct and erroneous solutions is determined by the equation $n_1 + n_2 = 2a$. When one is passed, the result is the same. It follows from this that the considered fragment of the circuit has the same properties and operates in the same way as the aforementioned known pre-detector linear addition circuit. The advantage is that it does not require the organization of a preliminary operation of phasing the added signals.

The part of the circuit outside the dotted line is a union of linear addition and auto selection. As shown in ([5]), it implements quasi-optimal processing, which is close in properties to optimal addition, but does not require complex-to-implement phasing and signal level control units. The circuit operates as follows. Using the amplitude detector (AD) and averager (AV) is determined by the current level of the signals in diversity branches. In the adder (Σ), their sum is formed with a weighting coefficient, the value of which must be equal to the value ([4]). All three received signals are compared, and the largest of them is selected (using the MAX block). The control signals generated by this unit, with the help of keys (K), are connected to the output through the OR circuit, either the output signals of one of the threshold devices (if there is an advantage in auto selection), or signal y, which corresponds to linear addition in properties.

Application of the channels described quasioptimal diversity signal transmission control algorithms can significantly simplify practical implementation of control.

3. Output

Pre-detector diversity combining always has advantages in noise immunity over post-detector diversity combining when considering only the linear properties of the signals. In the case of taking into account the nonlinear properties of analog signals, there are times when this advantage is lost. The use of nonlinear features of the combined diversity signals allows to increase the noise immunity of transmission and, with digital signals, to simplify the implementation of combination schemes.

References

[1] Sklyar B 2003 *Digital communication. Theoretical foundations and practical application* (M .: Izd. House "Williams") 1104 p
[2] Nemirovskij A S 1984 *Borba s zamiranijami pri peredache analogovyh signalov* (M: Radio i svjaz) 208 p
[3] Prokis J 2000 *Digital communication* (M: Radio and communication) 800 p
[4] Polushin P A 2011 *Methods of dealing with interference and distortion* (Saarbrucken, Germany: LAP LAMBERT Academic Publishing) 341 p
[5] Polushin P A and Samojlov A G 2007 *Izbytochnost signalov v radiosvjazj* (M:. Radiotechnika) 256 p
[6] Feer K 2000 *Besprovodnaja svjaz. Metody moduliatsii i rasshirenija spectra* (M: Radio i svjaz) 520 p
[7] Ipatova V P 2003 *Sistemy mobilnoj sviazi* (M: Gorjachaja linija-Telecom) 272 p
[8] Galkin V F 2007 *Tsyfrovaja mobilnaja svjaz* (M: Gorjachaja linija-Telecom) 432 p

[9] Polushin P A and Sinitsin D V 2014 Povyshenie kachestva i nadezhnosti peredachi signalov po kanalam svjazi i upravlenija *Tehnologija tekstilnoy promyshlennosti* **4(352)** pp 156-9
[10] Lathi B P and Ding Zhi 2009 *Modern Digital and Analog Communication Systems (Fourth ed.).* (Oxford University Press, Inc.) *p. 394.* ISBN 9780195331455.
[11] Anderson J B, Rusek F and Owall V 2013 Faster than Nyquist Signaling *Proceedings of the IEEE*

IOP Conf. Series: Materials Science and Engineering 1227 (2022) 012004 doi:10.1088/1757-899X/1227/1/012004

Methodology for assessing the aerodynamic imbalance of GTE impellers

E V Kudashov, M A Bolotov, I A Grachev and N D Pronichev

Samara National Research University, Institute of Engines and Power Plants, Samara, 443086, Russian Federation

E-mail: KEV-Fantom@yandex.ru

Abstract. Increased vibration during testing and operation of gas turbine engines may be caused by aerodynamic imbalance. As usual, aerodynamic imbalance is eliminated on stands simulating the operating conditions of the impellers. Existing stands have a high cost of testing or do not allow to make allowances for all the factors during the operation of the impellers. A review of models and techniques has shown the importance of considering operating parameters when calculating aerodynamic imbalance. The paper provides a methodology for calculating the aerodynamic imbalance of the impeller. The methodology is based on the use of finite element modeling (FEM). The proposed method for assessing the aerodynamic imbalance of the impellers makes it possible to take into account the rotational speed, the actual geometry of the blades based on the measurement results, as well as the assembly conditions and the operating conditions of the impeller. The technique was used through the example of a specially designed and manufactured impeller with geometric deflections. Theoretical and experimental studies have been carried out. The discrepancy between theoretical and experimental data did not exceed 15.5%. The developed model can be used at enterprises that manufacture gas turbine engines, gas compressor units, steam turbine plants and wind power plants.

1. Introduction

Dynamic balancing of the impellers is an integral part of the technological process of assembling the rotors of gas turbine aircraft engines. The rotation frequency at the stage of balancing of serial machines is 10...50% of the working one. The quality assessment of the balanced rotor of a gas turbine engine is carried out according to the interstate standard [1]. Dynamic balancing does not allow eliminating the aerodynamic imbalance of the rotor, since it does not take into account the real rotational speed, the environment where it operates and the factors. The aerodynamic imbalance of the impeller can be up to 20% of the total imbalance.

Aerodynamic imbalance arises due to the uneven gas flow through the passageway of the impeller. Mechanical and aerodynamic imbalances are similar in that during operation there is an unbalanced force vector pointed to the radial direction of the impeller. The above vectors can be positioned to each other quite randomly. Aerodynamic imbalance results from the following factors:

- the manufacturing error of blades and disks;
- the manufacturing tolerance of the installation of the blades in the disks and their pitches [5, 6];
- the spread in values of the blade airfoils deformations under the action of centrifugal and gas forces, resulting in a different change in the angles of profiles and pitches installation [6];

- the spread in values of the tensions over the contact surfaces of the blade shrouds [7] and other factors.

Mechanical imbalance of the impeller is eliminated using balancing stands by removing or adding corrective masses, and aerodynamic imbalance is eliminated using stands that simulate the operating conditions of the unit. At most enterprises, measures to eliminate the aerodynamic imbalance of the impellers are not carried out due to the high cost of the stand and the significant costs of testing. Another option is the use of acceleration benches, which allow the units to be accelerated to operating speeds. The use of acceleration benches does not allow to reduce the aerodynamic imbalance, since the factors acting during the operation of the impellers are not taken into account.

The invention [2] describes a method for balancing the aerodynamics of a vane wheel, based on an amplitude value estimation of the radial forces of the blades when they are selected to the scheme of their balance. The paper [3] presents a method of low-speed balancing of the mass of a high-speed blade rotor, which consists in balancing the rotor using a balancing machine. There is a way [4] of balancing the vane wheels of machines, which consists in the selection of blades with close chord angles. Their measurement is carried out in a certain section of the blade airfoil to the wheel plane, and they are installed in pairs in mutually opposite grooves of the disc. The described methods increase the labor intensity and cost of finishing work, and also do not take into account the temperature, pressure and composition of the gas air environment during operation. The paper [8] describes the importance of taking into account the total pressure and air inlet temperature $\left(P_{in}^{*}, T_{in}^{*}\right)$ when simulating the flow of a gas medium through a gas turbine engine. The operation of an aircraft engine is associated with a change in P_{in}^{*} and T_{in}^{*}, which change the forces acting on the blades and on the impeller as a whole.

Analytical and FEM do not have these disadvantages. As the study [9] shows, the aerodynamic imbalance of the impeller can be balanced by the method of mass correction for one of the operating modes. This is because the emerging force is directly proportional to the angular velocity of the rotor, and to compensate it, the centrifugal force of the correcting mass is used, which is proportional to the square of the angular velocity. The problem of aerodynamic imbalance also arises in wind turbines [10] as a result of deformation of the blades. The paper provides a forecast of the frequency and damping coefficient depending on the rotational speed of the blade.

Computer modeling makes it possible to improve the aerodynamic and aeroacoustic characteristics of the blade airfoils [11], as well as to predict the dynamic stability of the blade units of the GTE compressor [12]. This makes it possible to design the geometry of the blades for the required operating conditions at the design stage. The paper [13] presents an analytical model for calculating the dynamic and aerodynamic imbalance for the Vestas V80-2MW wind turbine. The article [14] investigates the influence of the main geometric parameters of the axial stage on the averaged and unsteady airloads acting on the rotor blades. Aerodynamic imbalance also affects the energy performance of engines [15].

A review of models and techniques has revealed the importance of considering operating factors when calculating aerodynamic imbalance. The paper proposes a method for assessing the aerodynamic imbalance of the impellers which takes into account the rotational speed, the actual geometry of the blades based on the measurement results, assembly conditions and operating conditions.

2. Methodology for assessing the aerodynamic imbalance of the impeller

2.1. Preparatory stage

The technique for assessing the aerodynamic imbalance of the impellers is presented in the form of a flowchart in figure 1.

Let us consider the application of the proposed technique on the example of a specially designed and manufactured impeller, taking into account its geometric deflections.

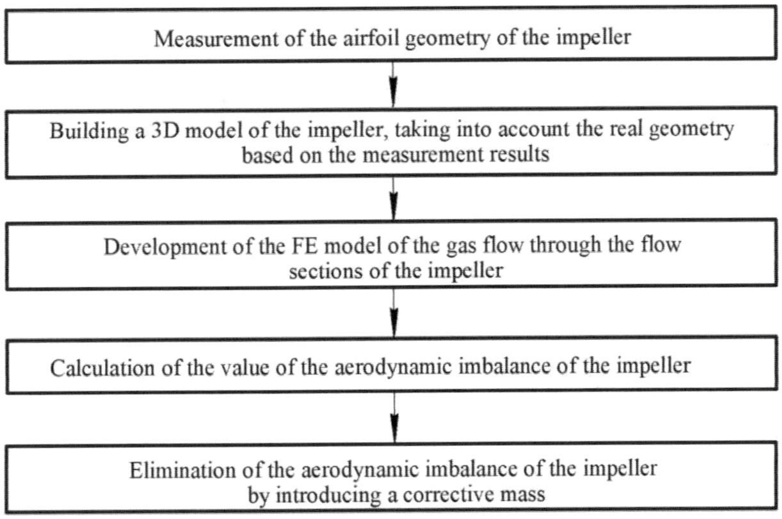

Figure 1. Block diagram of a technique for assessing the aerodynamic imbalance of the impeller.

At the preparatory stage, the geometry of the impeller blading was formed using the ANSYS software product with the aid of the Vista AFD and BladeGen modules. The initial parameters of the process were the characteristics of a mobile balancing machine used for small rotors: shaft speed $n = 5800\ rpm$; shaft power $N = 100\,W$; shaft torque $M_k = 0{,}17\ N \cdot m$; the outer diameter of the impeller $D = 84\ mm$. As a result of the calculation in the Vista AFD module, the geometry of the blade airfoil was formed and the following parameters of the impeller were determined: air flow rate $G = 0{,}195\ kg/\sec$ and its total pressure increase $P^* = 800\,Pa$, the number of blades was 7 pcs. The resulting geometry was named Impeller No. 1.

The calculated nominal geometry of impeller No. 1 has been revised. The blade airfoils were changed and the impeller was constructed in the BladeGen module. The impeller has small dimensions, and the aerodynamic imbalance is mainly significant for oversized fan impellers, therefore, the location of the two blades had to be profoundly changed relative to the nominal position. The changed parameters of impeller No. 2 relative to the nominal (No. 1) are presented in table 1.

Table 1. Changed parameters of impeller No. 2.

Blade, number	The angular position of the blade relative to the nominal	The angle of the blade relative to the nominal
1	5°	-10°
2	-51.43°	0°
3	-102.86°	0°
4	-159.29°	10°
5	-205.71°	0°
6	-257.14°	0°
7	-308.57°	0°

Geometric deflections in 1 and 4 of the impeller blades will cause flow separation and the appearance of unbalanced forces on them. Due to the opposite rotation of two blades relative to each other, the total projection of the force vector will be headed in one direction.

2.2. Measuring the flow path of the impeller

The measurement was carried out with a RangeVision PRO optical scanner using a rotating table. Impeller No. 2 was manufactured using FDM additive technology. The measurement of impeller No. 2

was carried out in order to control the accuracy of its manufacture and to verify the adequacy of the developed technique for assessing the aerodynamic imbalance. Based on the measurements obtained, a 3D model will be built, taking into account the geometric deviations identified during its measurement. Scanning was carried out under the operating conditions of the first zone to ensure the highest accuracy. Impeller No. 2 was measured in three positions with a discreteness of 15 degrees in automatic mode. The measurement process is shown in figure 2.

Removal of outlying values, noises and stitching of the resulting 72 scanned copies was carried out in ScanCenter NG. As a result of the final alignment, the estimated stitching accuracy of the impeller scans was $0.025\,mm$. The claimed measurement accuracy of the RangeVision PRO scanner in the first zone is $0.018\,mm$. The resulting polygonal model, shown in figure 3, was exported to *.stl format for further processing. Impeller measurement can also be carried out by the contact method using a coordinate measuring machine. In this case, the predetermined sections of the blades will be controlled. For convenience of further use, a parametrized 3D model can be created that changes its geometry according to the measured coordinates of the surface points. In the case of creating a parameterized 3D model, the next step of the methodology is not required.

Figure 2. Impeller measurements with an optical scanner.

Figure 3. The resulting polygonal model of the part after stitching.

2.3. Building a 3D model that takes into account real deviations obtained from measurements

The current stage is being implemented using the Siemens NX CAD system based on a scanned polygonal model. Since a solid 3D model is required for further steps, it was first converted to a sheet body. Small unscanned facet elements were interpolated with cubic splines, and on their basis the "patch" surfaces were built. Surfaces not involved in further calculations have been removed or replaced with planes or cylinders. The next step was to transform the surfaces into a solid 3D model of impeller No. 2 and build its flow path using a Boolean operation. The CAD model of the part from the NX system is saved in *.x_t format. Then their geometry is imported into the CAE system.

2.4. FEM of the gas flow through the passage volume of the impeller

The calculations were performed in the CFX module of the ANSYS software product. The following parameters were set as the initial parameters: air flow rate G, the magnitude of the total pressure increase behind the impeller p_q^*, shaft speed n, and the environmental parameters. The CFX module simulated the operation of impellers No. 1 and No. 2. The divergences of forces no more than $0.0001\,N$ along the Ox, Oy, and Oy axes between iterations were used as repeatability criteria. Figure 4 shows a solid 3D model of the impeller flow volume, and figure 5 shows the lines of gas flow through it.

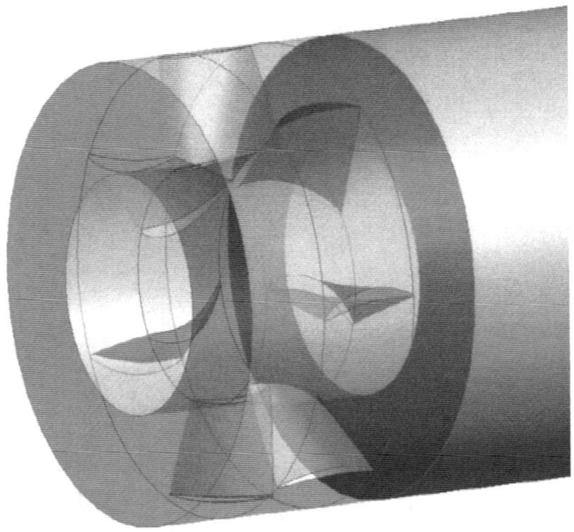

Figure 4. Solid 3D model of impeller No. 2.

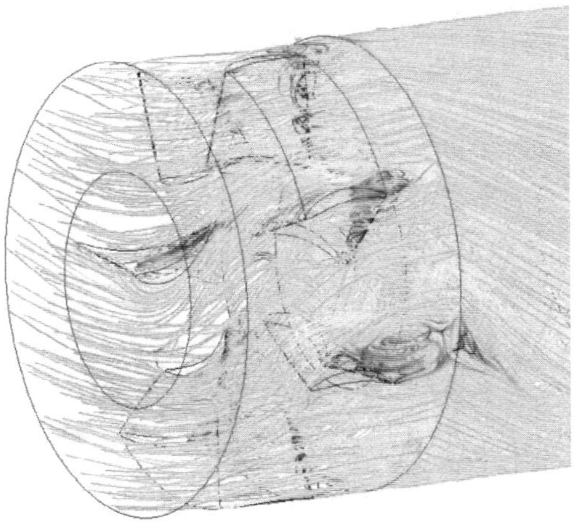

Figure 5. Lines of gas flow through the passage volume of impeller No. 2.

The result of the numerical calculation was: the projection of forces, co-directional to the coordinate axes of the rectangular coordinate system, the torque and power arising from the gas flow through the impeller.

2.5. Aerodynamic imbalance calculation
The calculation was carried out on the basis of the obtained projections of forces directed in the radial direction. To perform calculations of the obtained data, the MATLAB software package was used.

The aerodynamic imbalance D_a was calculated using the expression:

$$D_a = \frac{\sqrt{F_x^a + F_y^a}}{w^2}, \tag{1}$$

where F_x^a, F_x^a – the resulting projections of gas forces on the axis Ox and Oy, accordingly, resulting from the gas flow around the blades; w – rotational speed of the impeller.

Elimination of aerodynamic imbalance is carried out by adding a balancing weight, or by removing the weight from the ends of the impeller. The mass of the correction weight can be found by the formula:

$$m = \frac{D_a}{e}, \tag{2}$$

where e – is the position radius of the correction weight.

The angle of the corrective weight is determined from the formula:

$$\beta = ar\cos\frac{\left(\overline{F_x^a} + \overline{F_y^a}\right)\cdot\overline{F_x^a}}{\left|\left(\overline{F_x^a} + \overline{F_y^a}\right)\right|\cdot\left|\overline{F_x^a}\right|}, \tag{3}$$

where $\overline{F_i^a}$ – is the vector of forces directed along the axes Ox and Oy.

Next, the repeatability of the proposed model will be tested.

3. Model verification

The model was verified in the following sequence:

1) elimination of the mechanical component of the total imbalance of impeller No. 2;

2) determination and elimination of aerodynamic imbalance of impeller No. 2;
3) assessment of the discrepancy between the results obtained using the model and during the experiment.

The first step was to eliminate the mechanical imbalance. To eliminate the influence of the gas air environment, the impeller was balanced in a VV-1 high-vacuum chamber (figure 6). Balancing of impeller No. 2 was carried out using a developed mobile balancing stand, which includes a BalKom-4 device.

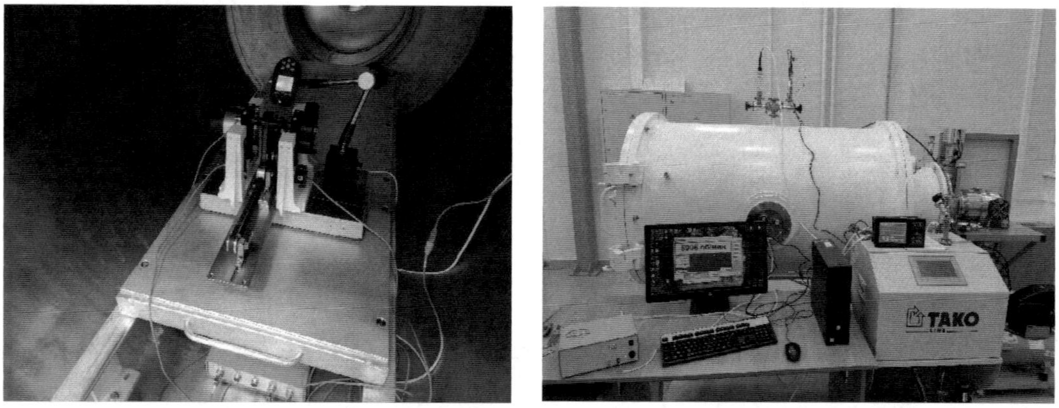

Figure 6. Mobile balancing stand installed in a high-vacuum chamber.

The balancing was carried out at a residual pressure of $1.3 \cdot 10^{-7}$ *bar* maximum and included 3 steps. Each step included removal of air, measuring, and normalizing the pressure inside the chamber to atmospheric. Elimination of the mechanical component of the imbalance was carried out by corrective weights installed in specially provided holes with a diameter of $2.5\,mm$, located on the separating radius of $18\,mm$. A weight of $0.15\,g$ was used as a test one. The electronic balance used in the experiment had a division value of $0.005\,g$. The residual mechanical imbalance of impeller No. 2 as a result of balancing in vacuum was $0.1\,g \cdot mm$, with a vibration level of $0.640\,mm/\sec$.

At the second stage, the mobile balancing stand was placed in atmospheric conditions, where after the aerodynamic imbalance was determined, as well as the position and weight of the corrective mass. After the introduction of the corrective mass, the residual imbalance was $0.1\,g \cdot mm$ with a residual vibration level of $0.613\,mm/\sec$.

The repeatability of the simulation results with the experimental data for impeller No. 2 was assessed by calculating the absolute deviations:

$$\Delta_a = P_m - P_e, \tag{4}$$

where P_m – is the parameter obtained as a result of modeling; P_e – is the parameter obtained during the experiment.

Relative deviations were determined from the formula:

$$\Delta_o = \frac{\Delta_a}{P_e}, \tag{5}$$

Table 2 presents the results obtained using the model and during the experiment for the calculated impellers.

IOP Conf. Series: Materials Science and Engineering 1227 (2022) 012004 doi:10.1088/1757-899X/1227/1/012004

Table 2. Theoretical and experimental data of the impeller.

Results	The magnitude of aerodynamic imbalance, $g \cdot mm$	Corrective weight mass, g	The position of the corrective weight, deg
FEM of impeller No. 1	0.056	0.0031	103.27
FEM of impeller No. 2	0.929	0.0516	290
Experiment carried out on impeller No. 2	1.1	0.059	326
Δ_a	0.155	0.0074	36
Δ_o %	15.5	12.5	11

The discrepancy between theoretical and experimental data when using the above test and measuring equipment was no more than 15.5%.

The discrepancy between the simulation and experimental results can be explained by the following reasons:

1) errors in measuring and creating models of impeller surfaces;

2) measurement errors of the BalKom-4 device, electronic scales;

3) the assumptions made when developing a model for predicting aerodynamic imbalance.

The developed model can be used at enterprises that manufacture gas turbine engines, gas compressor units, steam turbine plants and wind power plants.

Acknowledgements

This work is financed by the Federal State Budgetary Institution "Foundation for Assistance to Small Innovative Enterprises in Science and Technology" under the agreement 683ГУЦЭС8-D3/63745.

References

[1] Interstate standard "*Vibration. Balance quality requirements for rotors in a constant (rigid) state.*" (approved by Order of the Federal Agency for Technical Regulation and Metrology dated January 25, 2008 N 6-st)

[2] Suvorov L M 2007 Pat. of the Russian Federation No 2005133979/28 appl. 02.11.2005, publ. 27.06.2007

[3] Suvorov L M 2011 Pat. of the Russian Federation No 2009109011/28 appl. 11.03.2009, publ. 27.05.2011

[4] Zhavoronkov L A 1997 Pat. of the Russian Federation No 94015714/11 appl. 28.04.1994, publ. 20.06.1997

[5] Kiselev Yu V 1986 Investigation of the influence of the fan impeller inhomogeneity on the structure of vibration processes *Vibration strength and reliability of aircraft engines and systems* (Kuibyshev: KUAI) pp 76-81

[6] Idelson A M, Kuptsov A I 2006 Elastic deformation of fan blades as a factor, influencing the gas-dynamic unbalance *Samara Machine-Building Design Bureau»* vol 5 (Samara: SSAU), pp 234-7

[7] Sundukov A E, Sundukov E V, Plotnikov S M, Balakin A U, 2016 Some aspects of aerodynamic imbalance vibration-based diagnostic on bypass turbofan engine cosmologies *Materials of the reports of the international scientific and technical conference.* (Samara: SSAU) pp 231-2

[8] Idelson A M Modeling of aerodynamic imbalance on fan blades 2003 *Problems and prospects of engine building development : proceedings of the ISTC* (Samara: SSAU) pp 180-5

[9] Filimonikhin G B and Oleinichenko L S 2015 Investigation of the possibility of balancing the aerodynamic imbalance of the axial fan impeller by mass adjustment *Eastern european journal of advanced technologies* (Kivorograd) pp 30-5

[10] Owens B C, Resor B R, Griffith D T and Hurtado J E 2013 Impact of modeling approach on

flutter predictions for very largewind turbine blade designs *Annual Forum Proceedings - AHS International* (Mexico) pp 2766-76

[11] Kim J H, Ovgor B, Cha K H, Kim J H, Lee S and Kim K Y 2014 Optimization of the aerodynamic and aeroacoustic performance of an axial-flow fan *AIAA Journal* vol 52 (Incheon: Inha University) **52** 2032-44

[12] Kabannyk S, Stel'makh A L, Zinkovskii A and Savchenko K 2019 Method of prediction of dynamic stability of gas- turbine engine blade assemblies for subsonic flutter *26 Intermational congress on sound and vibration* (Montreal) hal-00756534

[13] Niebsch J, Ramlau R and Nguyen T T 2010 mass and aerodynamic imbalance estimates of wind turbines *Energies* (Altenbergerstr) **3** 696-710

[14] Kolenko G S and Laskin A S 2020 Mass and aerodynamic imbalance estimates of wind turbines *Scientific and technical bulletin of SPBGPU. Natural and engineering sciences* (St. Petersburg) **26** 15-28

[15] Korneev N V and Polyakova E V 2014 Aerodynamic disbalance of the turbocompressor as the reason of lowering of power indexes of internal combustion engine *Machinery engineering* (Moscow) **21** 51-7

IOP Conf. Series: Materials Science and Engineering 1227 (2022) 012005 doi:10.1088/1757-899X/1227/1/012005

The use of a computer modeling system for foundry processes in blank production

R A Vdovin[1], Iu V Komisar and A V Karpov

Samara University, 34, Moskovskoe sh., Samara, 443086, Russia

[1]E-mail: vdovin.ssau@gmail.com

Abstract. The article focuses on the analysis of the results of computer simulation of the technological process of casting the part "Stub". In the work, the elements of the gating-feeding system were calculated. Based on the calculated data, a digital model was built. An important stage of the work was the digital modeling of the technological process of casting the plug part in the ProCast CAE system. The simulation results were viewed and analyzed in the Visual-Viewer postprocessor module. An analysis of the simulation results made it possible to evaluate the efficiency of the gating system, determine the areas of turbulent flow, the nature of filling the zone with a high flow rate, possible places of mold erosion, places of formation of air pockets that can affect the occurrence of gas inclusions in the casting. The nature of the distribution of the temperature gradient in the casting, the time and uniformity of the solidification of the casting were also studied.

1. Introduction

Computer simulation and optimization of parts manufacturing technologies is an important direction in the development of software products for numerical engineering analysis. Computer analysis of foundry processes at the stage of digital design of casting technology (before castings are made) makes it possible to minimize possible miscalculations and errors that inevitably arise in the development process, reduce financial and time costs, increase efficiency, competitiveness, quality and reliability of the products being developed. Only computer simulation of the technology allows one to "look" inside the product, to see the nature of the processes occurring in it, to understand the causes of defects [1, 2].

In this work, the object of study was a thin-walled part - a plug, which is made of structural carbon steel grade St45L (figure 1).

The plug is installed along the outer ring into the motor housing and fastened with fasteners - studs.

The plug experiences axial and radial stresses, as well as alternating stresses from vibration loads and minor thermal loads.

From a design point of view, a part blank can be obtained both by stamping and casting.

Calculations of the unit cost of the manufactured part showed that the most effective method is investment casting, which allows to obtain a complex shape of the workpiece. Table 1 presents the main dimensions of allowances, overlaps and rounding radii necessary for designing a digital model of a part casting (figure 2).

In accordance with the technique of the technological process of modeling metal shrinkage in the manufacture of aerospace engineering parts, developed by the team of authors, implemented as a software module and intended for use in computer simulation systems for foundry processes, it was found that the total volumetric shrinkage for a workpiece of a plug part made of carbon steel mass 0.42 kg is 2%.

2. Description of research models

Table 1. General allowances for machining castings.

The largest overall dimension of the casting, mm	Allowances per side, mm[a]		
	Accuracy classes		
	LT1 LT2 LT3	LT4 LT5	LT6 LT7
Up to 60	0.5	2.0	2.5
From 60 to 100	0.5	2.0	3.0
From 100 to 160	0.5	2.5	3.5
From 160 to 250	0.7	2.5	4.0
From 250 to 400	1.0	2.5	4.0
From 400 to 630	1.5	3.0	5.0

[a] The smallest machining allowance must be greater than the tolerance.

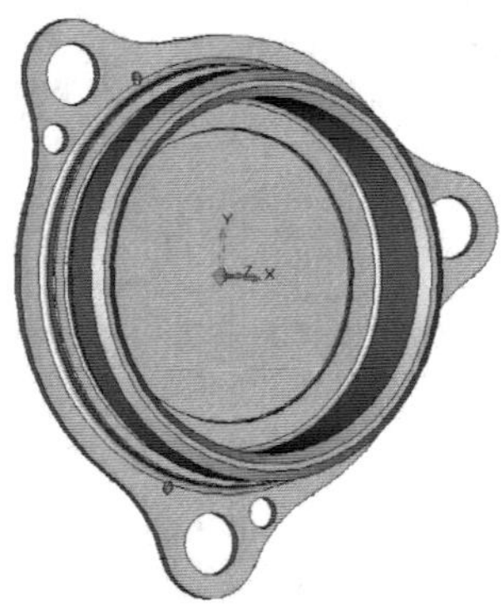

Figure 1. 3D model of the plug part.

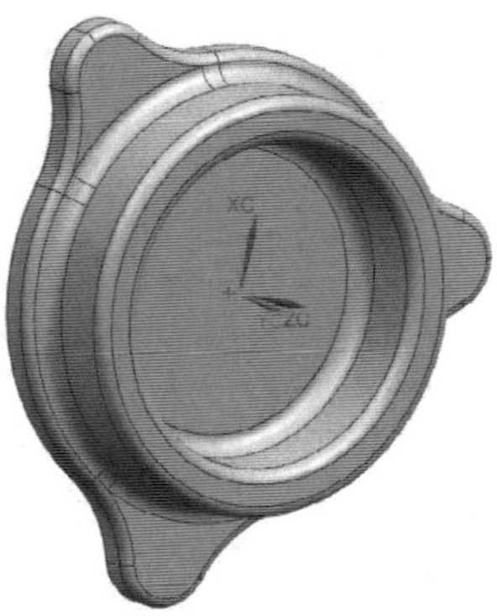

Figure 2. 3D model of the casting of the plug part.

For further analysis of the casting process and obtaining a casting of a stub part, it is necessary to calculate the elements of the gating-feeding system.

Pouring molds of small and medium-sized castings on conveyor and automatic lines, i.e. in serial and mass production, is carried out from rotary buckets with a capacity from 100 to 800 kg.

In this case, the consumption of steel from the ladle is controlled by the angle of its inclination. The limiting element of a closed or tapering system of ladle-gating channels-mold cavity are feeders with the smallest cross section.

In accordance with the methodology of B. Ovann, the design of the gating-feeding system, shown in Figure 3, was calculated and designed.

An important stage of the work performed was the digital modeling of the technological process of casting a plug part in the CAE software product ProCast.

The simulation results were analyzed in the Visual-Viewer postprocessor.

During the analysis of the process of pouring an alloy into a casting block, it is possible to determine the nature of the filling, evaluate the efficiency of the gating system, determine areas of turbulent flow, zones with high flow rates, possible areas of erosion of the mold or core, places of formation of air pockets that can affect the occurrence of gas inclusions in casting [3 – 5].

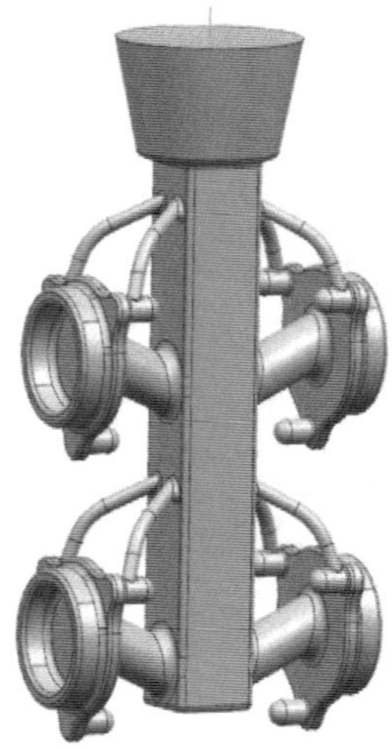

Figure 3. Computer model of the casting block.

3. Results and discussion

Based on the picture of the mold filling hydraulics (Figure 4), we can conclude that the optimal metal pouring speed was selected (it is 0.9 m/s), since the metal level in the sprue bowl is kept at a constant level, its flow occurs without unnecessary bursts and braking.

The dynamics of filling the casting block with metal indicates the optimal design of the gating-feeding system. This conclusion follows their calm and uniform nature of filling, as well as the correct and timely operation of the structural elements of the gating-feeding system (feeders, risers, risers, etc.).

The crystallization process starts from the bottom of the casting - these are the risers and the thinnest walls of the casting (Figure 5). The solidification of the casting occurs very quickly, because all the elements of the casting are thin-walled.

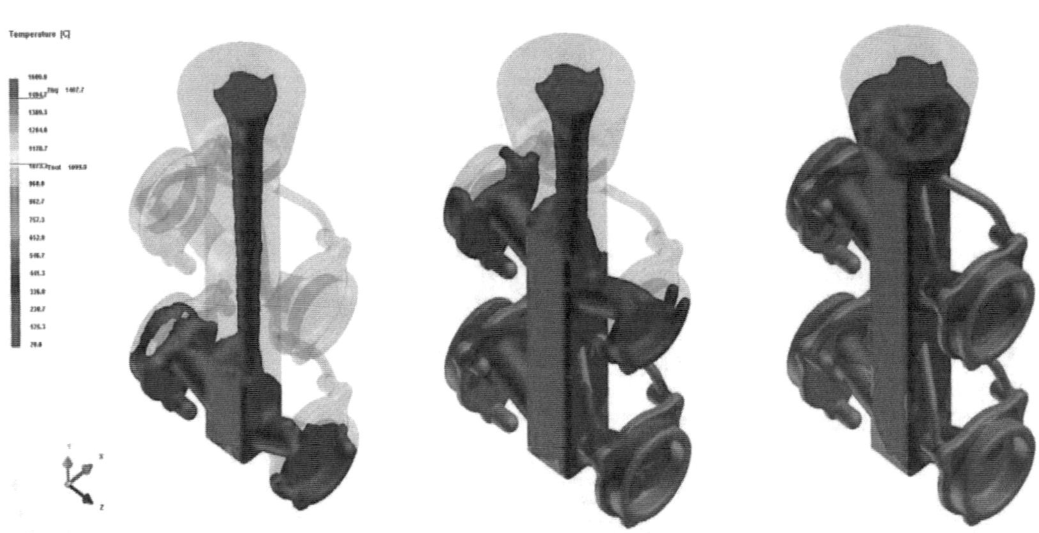

Figure 4. Pattern of filling the casting block with liquid metal at different points in time.

IOP Conf. Series: Materials Science and Engineering 1227 (2022) 012005 doi:10.1088/1757-899X/1227/1/012005

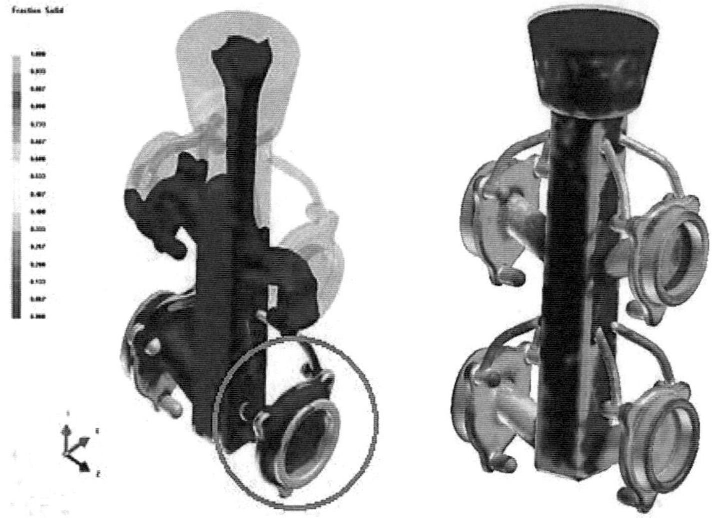

a *b*

Figure 5. Crystallization of the casting of the plug part:
a – at the initial moment; b - before complete hardening.

Crystallization of the central part of the casting of the stub part occurs quite dynamically and evenly (Figure 6). With a similar nature of solidification, fine looseness (macroporosity) may occur, which is associated with the phenomenon of volumetric crystallization and, as a result, volumetric shrinkage.

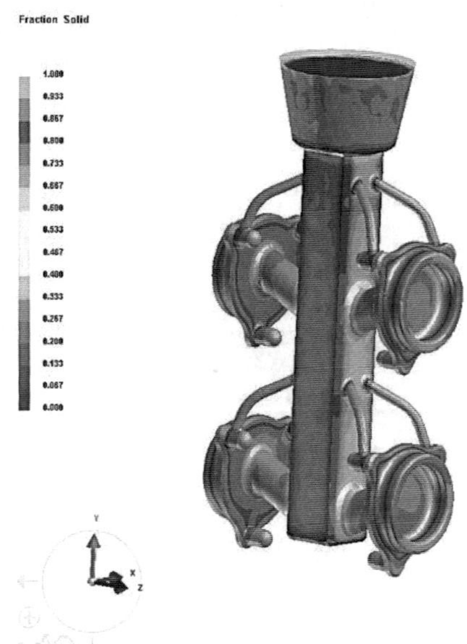

Figure 6. Pattern of X-ray control for predicting defects in the stub part.

The Chvorinov's Thermal Modulus parameter allows you to evaluate the geometry of the casting and the nature of its thermal layer (Figure 7). The higher the value of the thermal modulus, measured in centimeters, the longer the hardening time of this area. This module is calculated as the volume of the current node divided by the cross-sectional area. It is often used to evaluate feed efficiency gains and evaluate the crystallization direction of a casting. Ideally, the modulus at the bottom of the casting should have a minimum value, gradually increasing towards the top of the casting. Profit should have the highest value of the thermal modulus in order to have time to impregnate all the lower parts of the casting. The casting zones have practically the same modulus values. Crystallization is quite fast. This phenomenon contributes to the appearance of small ripples in this zone [6–10].

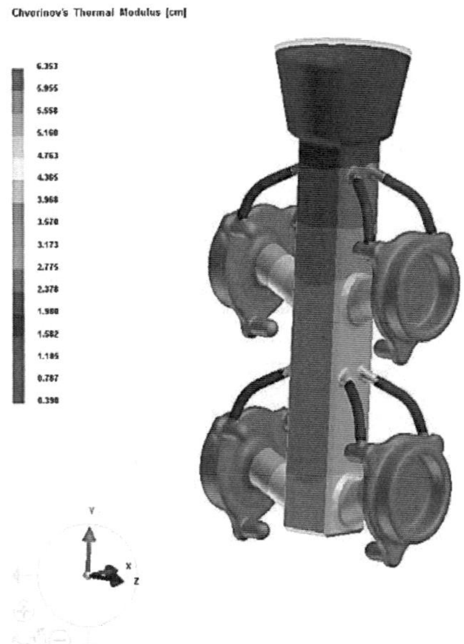

Figure 7. Thermal modulus evaluation pattern.

Hot nodes (Figure 8a) show the solidification time of the zones in the casting; thermal units that hardened last and where shrinkage is possible. These are places where shrinkage porosity appears as a result of delayed solidification - slower than in neighboring casting areas.

Porosity (Figure 8b) in the casting itself is within the normal range (up to 3%), but in the feeder and riser it is above the norm.

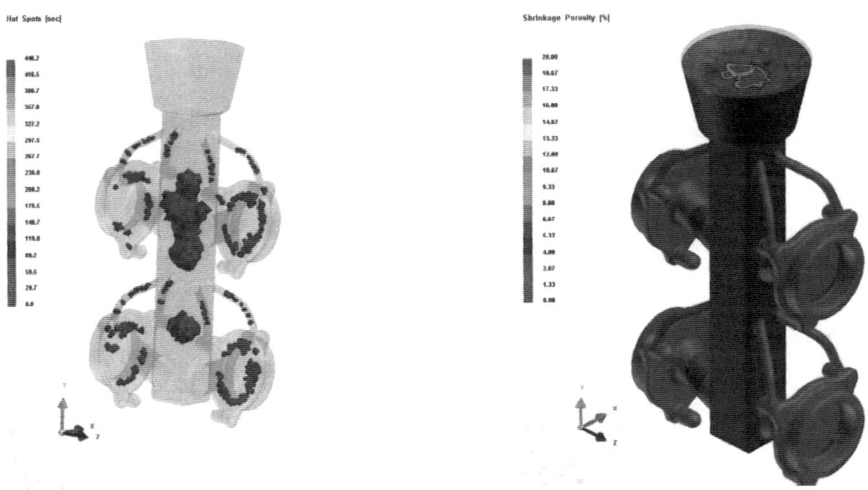

a *b*

Figure 8. Porosity observation pattern at the macro- and microlevels:
a – formation of hot knots; b – formation of porosity.

The Solidification Time parameter allows you to evaluate the time and uniformity of solidification of the casting. Figure 9 shows that the fastest places of solidification are in narrow sections (uplifts).

After that, you can observe the uniform hardening of the castings themselves, after the feeders and the riser.

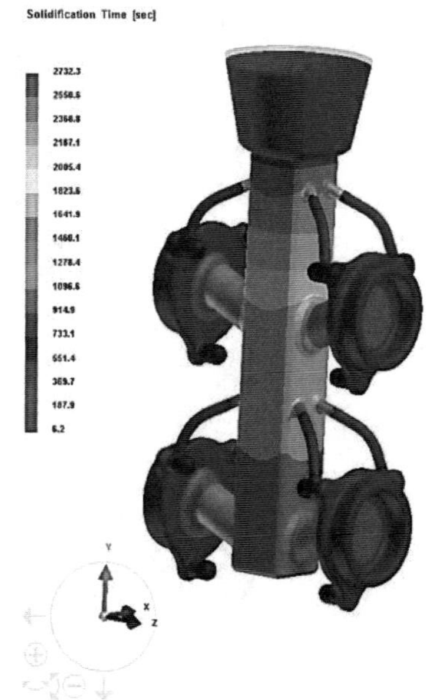

Figure 9. Casting solidification time.

With the help of the ProCast computer modeling system for foundry processes, it is possible to carry out particle tracing analysis to evaluate the nature of the filling, i.e. view the problem of getting foreign particles, molding components of the mixture or shell elements into the LPS, where they can be carried out by the metal flow during the pouring process (Figure 10). The program forcibly delivers particles to the metal supply zone after a specified time interval and the calculation of the movement of these particles inside the casting block is carried out. As a result, it can be seen that the maximum accumulation of particles occurs in the gate bowl and risers, to a lesser extent, directly, in the casting. However, if there are turbulences during movement (a kind of tangle of tracers), then the probability of accumulation of particles in this area increases. In this case, the particle motion trajectory is relatively directed – from top to bottom [11, 12].

4. Conclusions

Summarizing, it can be noted that the results of a computer analysis of the technological process of casting a plug part made it possible to evaluate the mechanism for obtaining a casting, the structure and causes of the formation of casting defects, on the basis of which comments were given on eliminating defects. The use of computer simulation systems for foundry processes makes it possible to reduce financial, material and time resources for the production of parts.

Thus, the ProCast software package, applicable to various casting processes and having a complex mathematical apparatus, adequately describes the physics of foundry processes and allows you to fully analyze the technology, taking into account the key features of foundry production. The flexibility of this package, combined with the hands-on experience of its developers, ensures that ProCast will meet any new challenges that come with changing technologies in the enterprise.

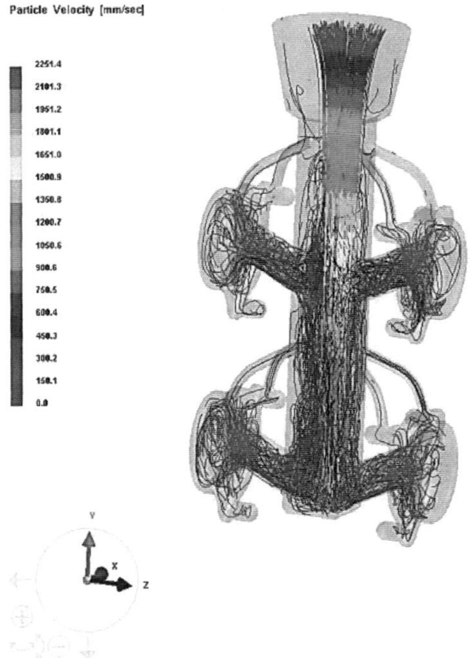

Figure 10. Tracing pattern of foreign particles in a casting.

Acknowledgment
The work was performed within the framework of a grant provided by the Innovation Fund of the Samara Region (IIF SO) under contract No. 1/1-IP/2021 dated November 08, 2021.

References

[1] Szeliga D P 2015 The unidirectional crystallization of metals and alloys (turbine blades) *Rzeszow University of Technology* **1** 45
[2] Vdovin R A 2019 Investigation of the linear shrinkage of photopolymer material during the technological preparation of the production of engine parts *Journal of Physics: Conference Series* **1679 (4)** 4
[3] Khaimovich I N and Demianinko E G 2015 *Mathematical material and process simulation* (Samara: SSAU) p 82
[4] Szeliga D P 2014 Modeling of directional solidification of columnar grain structure in CMSX–4 nickel-based superalloy *Materials Engineering and Performance* **23 (3)** 1088-95
[5] Vdovin R A 2020 Study of the stability of the geometric dimensions of wax models of GTE blades using rapid prototyping technology *IOP Conference Series: Materials Science and Engineering* **919 (2)** 5
[6] Takeshi N 2014 Development of CMC turbine parts for aero engines *Engineering Review* **47 (1)** 29-32
[7] Vieira L F, Paggiand R A and Salmoria G V 2012 Innovative developments in virtual and physical prototyping *Proceeding* 385-8
[8] Chubin Y 2013 Competitive grain growth mechanism in three dimensions during directional solidification of a nickel-based superalloy *Journal of Alloys and Compounds* **578** 557-84
[9] Dai H J, Dong H B and D'Souza N 2011 Grain selection in spiral selectors during investment casting of single-crystal components: Parts II Numerical modeling *Metallurgical and Materials Transactions* **42 (A)** 3439-46

[10] Vdovin R A 2020 Analysis of the process of shaping the blanks of the GTE turbine working blades *Journal of Physics: Conference Series* **1679 (4)** 6
[11] Bolotov M A, Pechenin V A and Ruzanov N V 2016 Uncertainties in measuring the compressor-blade profile in a gas-turbine engine *Russian Engineering Research* **36 (12)** 1058-65
[12] Pechenin V A, Bolotov M A and Rusanov N V 2014 Method of evaluation of profile form and shaped surfaces with application of wavelets *Research Journal of Applied Sciences* **9 (11)** 820-4

IOP Conf. Series: Materials Science and Engineering 1227 (2022) 012006 doi:10.1088/1757-899X/1227/1/012006

Mathematical Model Design for a 7-DOF robot and 6-DOF comparison simulation

José Luis Ordoñez-Avila[1], Roger Midence[1] and Islam A Magomedov[2]

[1] FI, Universidad Tecnológica Centroamericana (UNITEC), San Pedro Sula, Honduras
[2] Kadyrov Chechen State University, Grozny, Russia

E-mail: jlordoenz@unitec.edu, ismwork@mail.ru

Abstract. Currently, serial manipulators are the boom in the industry. Most companies are looking for solutions that improve production processes and that are human collaborative. They seek to simulate the skills that humans perform. Nowadays there is a constant question of comparison between a redundant robot of 7-DOF and one that meets the necessary joints to perform complex jobs useful in the industry as is the 6-DOF. In this paper we design a mathematical model that describes the forwards kinematics of the 7-DOF manipulator using the famous D-H convention and determine if the kinematic chain of the manipulators interferes in the performance of their work and thus reach a comparison with its predecessor of 6-DOF in the industry by means of a simulation in CoppeliaSim. As the main results the 6-DOF is more useful for repetitive production assignments and the 7-DOF for human collaboration.

1. Introduction

This article shows first the development of a mathematical model developed in Matlab and then the simulation in Coppeliasim of the manipulator of seven degrees of freedom (7-DOF) in comparison with one of 6 degrees of freedom (6-DOF). For this, references were consulted about the last 5 years from three libraries (IEEExplore, IOPScience and ACM Digital Library). Figure 1 shows the line time of the investigations that have as title "Seven Degrees of Freedom" in the last two years this topic has increase its interest [1-10].

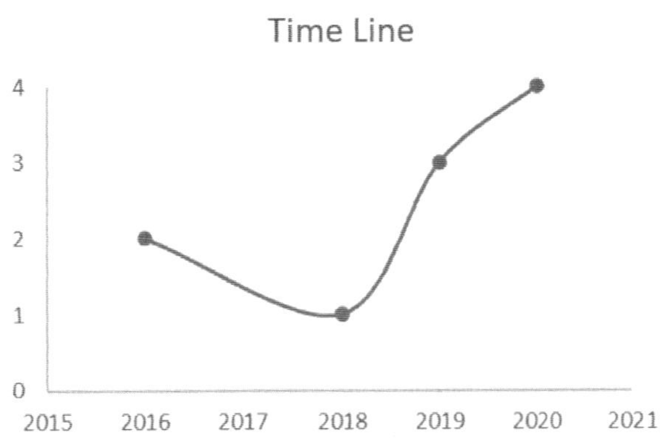

Figure 1. Five years 7-DOF research timeline.

Within this review of references, 10 articles were found, which were divided into three main topics. This highlights the development in Matlab of the angular displacements of the actuators through genetic algorithms, through a graphical interface and the construction of the manipulator [3]. Therefore, the articles describing applications of 7-DOF [1-3] were placed in the first topic. The study of the kinematics and dynamics of these manipulators is the most developed topic, proposing templates and methods based on 40% of the Denavit Hartenberg (D-H) algorithm [4-8]. An example is the proposed method for redundant manipulator configuration, obtaining a complete analytical inverse kinematics equation in the position space, which eliminates the Jacobian matrix [6]. Finally, 20% of the articles make comparisons between robots with 7-DOF and those with 6-DOF [9-10]. Matlab for both applications, kinematics and dynamics is the most used software with 70%, followed by ROS with 20% to simulate the models developed in Matlab as shown in figure 2.

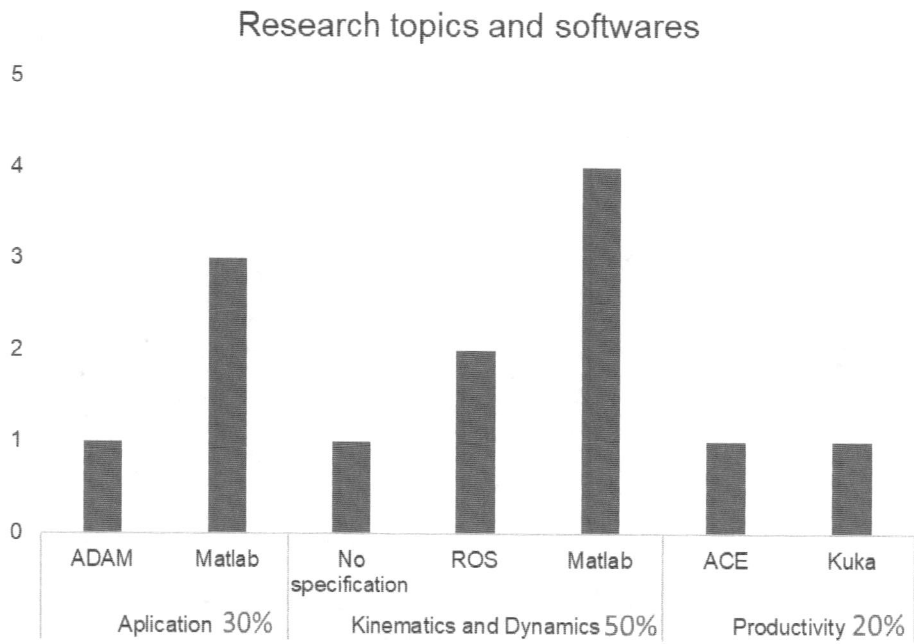

Figure 2. Five years 7-DOF research topics and software.

These papers are not enough to show the advantages and disadvantages of these manipulators. So, the objective of this research is to compare 7-DOF and 6-DOF manipulators in industrial settings to understand their characteristics. This comparison also includes a mathematical model of forward kinematics.

2. Methodology

In order to develop a research project in robotics in an effective way, a methodical study is required to evaluate the risks present in the different stages of development and take into account the implementation of solutions for the delivery of an efficient work that meets the objectives set. To achieve this objective, it is determined to use a spiral methodology in this article, a methodology that helps to consider the risks presented in the different iterations, a methodology that solves the risks that arise, it is a methodology that manages to implement the proposed solutions achieving a successful conclusion of the project if it is applied in the best way.

2.1. Kinematic diagram design

This stage focuses on using the D-H convention to create the kinematic diagram of the Kuka manipulator based on the joint structure and structure of its links, a sequence of cylindrical representations is determined to denote the degrees of freedom in revolute, in case the degree of freedom was prismatic, it would be exposed as a cube. The kinematic diagram of the degrees of freedom is shown in figure 3a, representing the rotation of each joint with the theta parameter.

2.2. Implementation of the D-H algorithm

Having made a kinematic diagram of the joints and solved the risks of the previous stage, we proceeded to perform the series of steps of the D-H algorithm where it is instructed how to obtain the position of the end effector with respect to its base corresponding to the relative positions of the rigid body of its links that 0ξ7 resulting in a more concrete diagram that includes the coordinate systems. The D-H algorithm has high degree of accuracy, although it has two constraints that refers where xi axis intersects zi-1 and xi axis is perpendicular to zi-1, the consequence of this constraint refers in the occasions that the coordinate frames of the links are not quite in the physical links of the manipulator. Another consequence is the robot to be placed in a specific configuration known as the zero-angle configuration, which all angles must be placed on 0 degrees [11]. It is necessary to take into account different aspects of the dimensional structure of the manipulator links, where the result of the mathematical model interferes due to its working space. Table 1 shows the D-H parameters of the working space of the Kuka manipulator.

Table 1. D-H Parameters 7-DOF manipulator.

i	θi	*di*	*ai*	*αi*
1	$\theta 1^*$	d1	0	-90
2	$\theta 2^*$	0	0	90
3	$\theta 3^*$	d2	0	-90
4	$\theta 4^*$	0	0	90
5	$\theta 5^*$	d3	0	-90
6	$\theta 6^*$	0	0	90
7	$\theta 7^*$	d4	0	0

2.3. Matrix resolution

In this stage we have the help of Matlab software for its great capacity to solve the linear algebra, in this stage a code is generated where all the individual matrices are obtained based on the generalized matrix (1) obtaining the matrices 0T1, 1T2 .... Having all the individual matrices, we proceed to obtain the matrix 0T7 by solving (2) and thus obtain the mathematical model of direct kinematics with the matrix named in Tt, which is the same as 0T7 in the literature. (3) shows the last matrix without evaluation and without being simplified. Figure 3b shows the Matlab simulation for the manipulator.

$$ {}^{0}\boldsymbol{T}_{7} = {}^{0}\boldsymbol{T}_{1}\,{}^{1}\boldsymbol{T}_{2}\,{}^{2}\boldsymbol{T}_{3}\,{}^{3}\boldsymbol{T}_{4}\,{}^{4}\boldsymbol{T}_{5}\,{}^{5}\boldsymbol{T}_{6}\,{}^{6}\boldsymbol{T}_{7} \tag{1} $$

$$ \begin{matrix} S(th_7)\sigma_5 - C(th_7)\sigma_4 & S(th_7)\sigma_4 + C(th_7)\sigma_5 & \sigma_3 & L_4\sigma_3 + L_3\sigma_{23} + L_2\sigma_{24}S(th_2) \\ C(th_7)\sigma_2 - S(th_7)\sigma_6 & -S(th_7)\sigma_2 - C(th_7)\sigma_6 & \sigma_1 & L_4\sigma_1 - L_3\sigma_{10} + L_2S(th_2)\sigma_{23} \\ C(th_7)\sigma_7 - S(th_7)\sigma_9 & -C(th_7)\sigma_9 - S(th_7)\sigma_7 & \sigma_8 & L_1 + L_3\sigma_{14} + L_4\sigma_8 + L_2C(th_2) \\ 0 & 0 & 0 & 1 \end{matrix} \tag{2} $$

X=*L4*(C(th6)*(S(th4)*(S(th3)*S(th1 - pi/4) - C(th2)*C(th3)*C(th1 - pi/4)) + C(th4)*C(th1 - pi/4)*S(th2)) - S(th6)*(C(th5)*(C(th4)*(S(th3)*S(th1 - pi/4) - C(th2)*C(th3)*C(th1 - pi/4)) - C(th1 - pi/4)*S(th2)*S(th4)) + S(th5)*(C(th3)*S(th1 - pi/4) + C(th2)*C(th1 - pi/4)*S(th3)))) +*

L3*(S(th4)*(S(th3)*S(th1 - pi/4) - C(th2)*C(th3)*C(th1 - pi/4)) + C(th4)*C(th1 - pi/4)*S(th2)) + L2*C(th1 - pi/4)*S(th2)

Y=L4*(S(th6)*(C(th5)*(C(th4)*(C(th1 - pi/4)*S(th3) + C(th2)*C(th3)*S(th1 - pi/4)) + S(th2)*S(th4)*S(th1 - pi/4)) + S(th5)*(C(th3)*C(th1 - pi/4) - C(th2)*S(th3)*S(th1 - pi/4))) - C(th6)*(S(th4)*(C(th1 - pi/4)*S(th3) + C(th2)*C(th3)*S(th1 - pi/4)) - C(th4)*S(th2)*S(th1 - pi/4))) - L3*(S(th4)*(C(th1 - pi/4)*S(th3) + C(th2)*C(th3)*S(th1 - pi/4)) - C(th4)*S(th2)*S(th1 - pi/4)) + L2*S(th2)*S(th1 - pi/4)

Z=L1 + L3*(C(th2)*C(th4) + C(th3)*S(th2)*S(th4)) + L4*(S(th6)*(C(th5)*(C(th2)*S(th4) - C(th3)*C(th4)*S(th2)) + S(th2)*S(th3)*S(th5)) + C(th6)*(C(th2)*cos(th4) + C(th3)*S(th2)*S(th4))) + L2*C(th2)

(3)

2.4. Implementation of the D-H algorithm

At this stage it is proposed to simulate the mathematical model in Matlab, the programming codes establish to replace the dimensions of the links in table 1, and thus obtain a numerical result when validating the prototype of the mathematical model in (3). When simulating the mathematical model in Matlab we will obtain the coordinates of the final effector based on the joints configurations where its workspace is defined. The manipulators were also simulated in an industrial environment using the CoppeliaSim software. This simulation will be done to achieve the results of four tests (avoiding obstacles, different orientations, velocity, acceleration, and productivity).

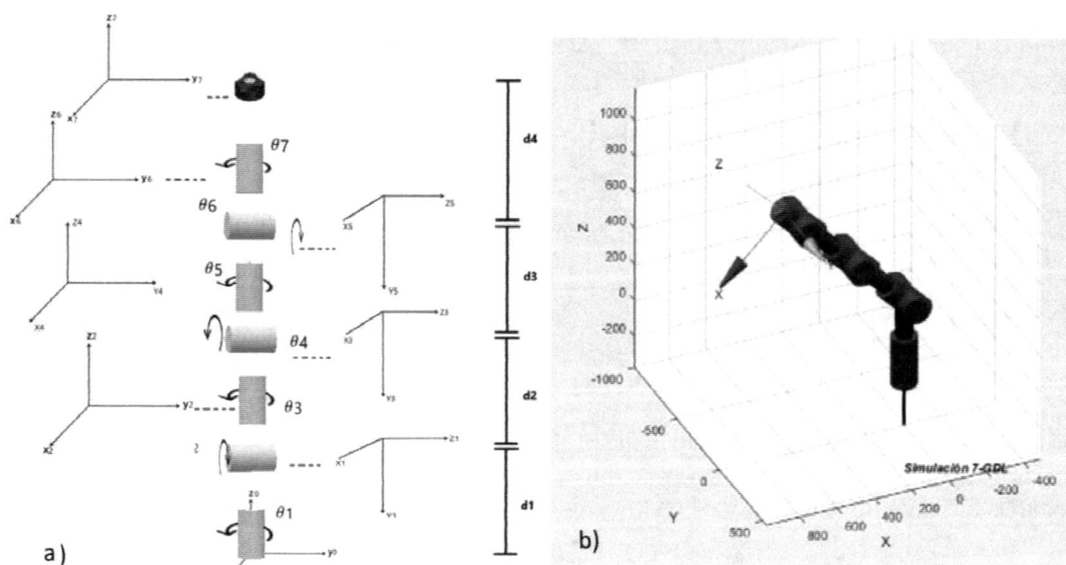

Figure 3. 7-DOF structure diagram Matlab simulation.

3. Results

In order to give a result to the design of the mathematical model we used Matlab with Peter Corke's Toolbox. Figure 3b shows a change in its second and third joint of 45 degrees, giving as valid the joint structure of the mathematical model. The proposed diagram was simulated in Matlab with the D-H algorithm where the mathematical model was validated. The accuracy of the mathematical model to obtain the Cartesian coordinates was validated using the CoppeliaSim.

This validation was carried out three times to best validate the mathematical model. Once the mathematical model was validated, the Kuka arm was compared to the ABB manipulator. The test consists of testing them in industrial pick and place processes, where they will have to pass a series of comparison challenges, the mathematical model will be used to find a point that the end effector of both manipulators be reachable at the deposit point in the process, the mathematical model validated the coordinates.

The process consists of choosing objects of different colors. The objects will pass through a conveyor belt, a sensor will detect if it is the selected color, and the robotic arm will come into action and transfer it to the second belt. A proximity sensor is located on this second belt. When the sensor receives information, the second belt will begin to move the object.

3.1. Avoiding Obstacles Test

By placing an obstacle in the same position for both manipulators as shown in figure 4. It is noted that the 6-DOF serial manipulator suffers collisions at its joint number 5. The 7-DOF manipulator can avoid this obstacle, due to its extra degree of freedom.

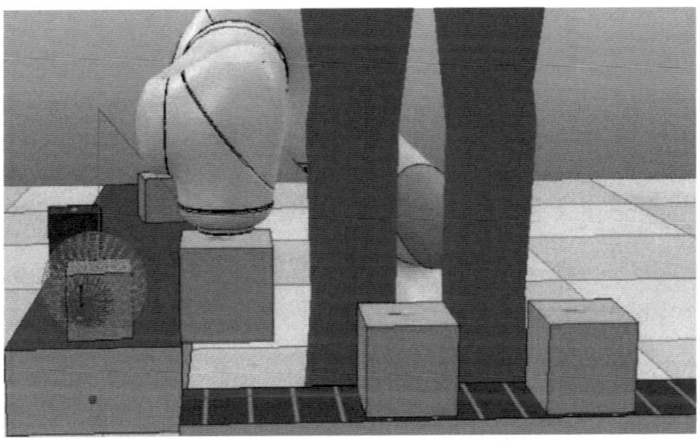

Figure 4. KUKA (7-DOF) avoiding obstacles test.

3.2. Different Orientations Test

In this test the orientation of the robots is changed to take the objects in the same position. The KUKA robot succeeded in replacing its previous inverse kinematics with a new joint configuration that avoided the obstacle and gave the human even more workspace (figure 5a). This 7-DOF manipulator provides an extra solution in scenarios where the final orientation of the object needs to be changed. In contrast, the ABB robot in the simulation failed to dodge the obstacle, nor did it achieve a change in inverse kinematics to achieve new joint configurations and reach the goal (figure 5b).

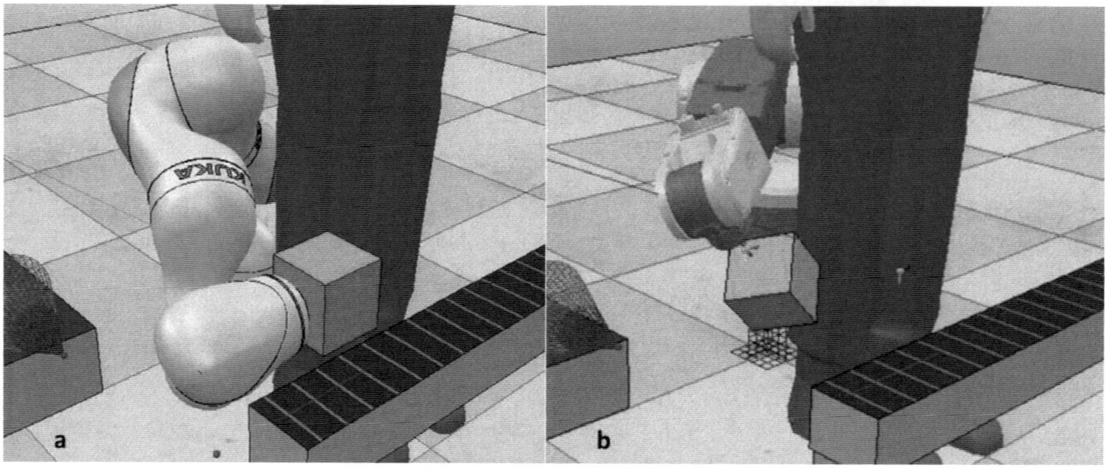

Figure 5. a) KUKA (7-DOF) different orientation test b) ABB (6-DOF) different orientation test.

3.3. Velocity and Acceleration Test

Three iterations were performed for each 0.8935m manipulator. The difference between the two is 0.00012s in favor of ABB. In terms of acceleration, ABB also has a small difference in its favor. Figure 6 shows the behavior of ABB in the iteration with the highest speed 0.04593m / s in 19.45s. Table 2 shows the average of the velocities and accelerations after the three iterations.

3.4. Productivity Test

The productivity of both manipulators was evaluated by conducting a real-time test of the industrial process with different speeds. Three tests were carried out, at 25%, 45% and 65% of their joint velocities, during a period of 20 minutes. The ABB robot was the most productive at each of the speeds. The average of the iterations in each speed is shown in table 2.

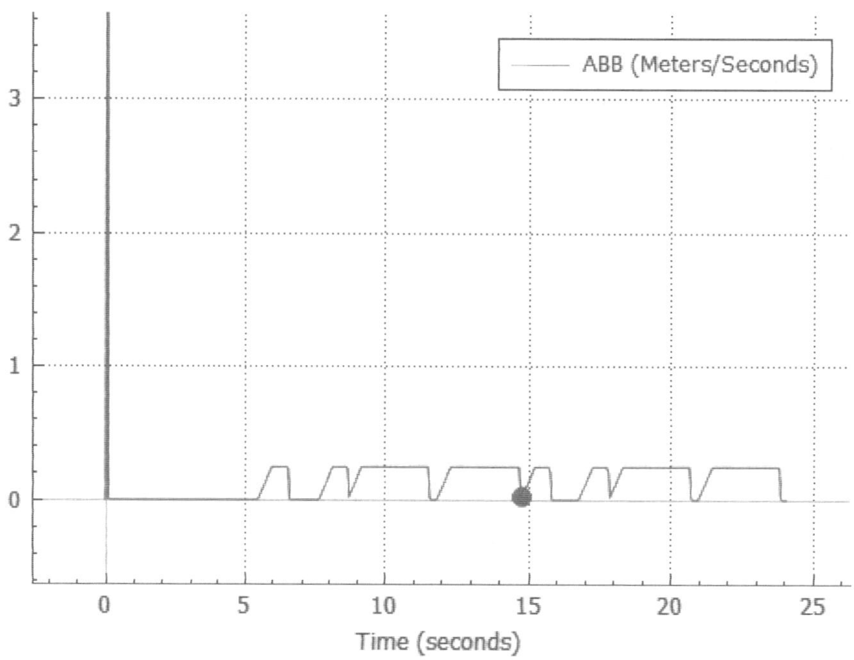

Figure 6. ABB (6-DOF) velocity and acceleration test.

Table 2. Test results.

Tests	KUKA 7-DOF	ABB 6-DOF
Avoiding Obstacles	Yes	No
Different Orientations	2	1
Velocity (m/s)	0.04517	0.04529
Acceleration (m/s^2)	0.00218	0.00229
Productivity (25%)	34	58
Productivity (45%)	207	297
Productivity (65%)	270	346

4. Conclusions

In this article, results were achieved with respect to a mathematical model that solved the forward kinematics of the Kuka manipulator using the D-H algorithm that was validated in certain ways with the help of simulators that met the needs of the research, such as CoppeliaSim conformed the industrial process and helped in the validation of the accuracy of the forward kinematics data provided by the mathematical model. Matlab was of great help in solving the homogeneous transformation matrices generated by the algorithm, the whole validation process was a success with the simulators and the algorithm.

Verifying the analysis made, it was possible to understand the behavior of the redundant manipulators with respect to their kinematics by understanding the motion of rigid bodies. The analysis based on the study made by [12] was of great help due to the analogy made of the human arm with the robotic arm. It was also possible to understand the kinematics thanks to the studies made by [11] where its being examples of different manipulators with different degrees of freedom using the Toolbox elaborated by the author of the book.

A concrete comparison was made by average of several tests with respect to joint movements and production speeds of a redundant manipulator of 7-DOF in contrast with 6-DOF. The results shows that the 6-DOF is more useful for repetitively process where humans have no interaction with the manipulator and no obstacle avoiding is needed. This 6-DOF has more speed and more acceleration that increase its productivity. The 7-DOF is more useful to collaborate with humans in production process and in process where is very difficult to pick an object due to orientation issues.

References

[1] M Kühne *et al.*, 2018 An MR-Compatible Haptic Interface With Seven Degrees of Freedom, *IEEE/ASME Transactions on Mechatronics*, **23(2)** pp. 624-635, abr., doi: 10.1109/TMECH.2018.2806440.

[2] L Xuan, S Peng, T Guan and N Li, 2019 Research on the Motion and Dynamics of Biomimetic Manipulator with Seven Degrees of Freedom, *International Conference on Mechatronics and Automation*, pp. 1830-1835.

[3] A Sánchez *et al.*, 2016 Design and construction of an anthropomorphic robotic arm of seven degrees of freedom with kinematic and dynamic analysis based on genetic algorithms, *IEEE International Conference on Automatica* (ICA-ACCA), oct. 2016, pp. 1-8. doi: 10.1109/ICA-ACCA.2016.7778502.

[4] Q Zhang, Y Shen, S Li, F Han and Z Wang, 2019 A New Analytical Inverse Kinematics Model for Seven Degrees of Freedom Redundant Manipulators, *International Conference on Intelligent Computing, Automation and Systems* (ICICAS), pp. 692-696. doi:

10.1109/ICICAS48597.2019.00150.
[5] Y Gloumakov, A J Spiers and A M Dollar, 2020 Dimensionality Reduction and Motion Clustering During Activities of Daily Living: Three-, Four-, and Seven-Degree-of-Freedom Arm Movements, *IEEE Transactions on Neural Systems and Rehabilitation Engineering* **28(12)**, pp. 2826-2836, dic. 2020, doi: 10.1109/TNSRE.2020.3040522.
[6] J Guo, Y Xu, C Huang, X Zhu and C Shao, 2020 An Analytical Solution to Inverse Kinematics of Seven Degree-of-freedom Redundant Manipulator *IEEE 9th Joint International Information Technology and Artificial Intelligence Conference* (ITAIC), **9**, pp. 2040-2050. doi: 10.1109/ITAIC49862.2020.9338865.
[7] J Wu and X Ren, 2020 Inverse kinematics solution and motion simulation of seven-degree-of-freedom ascending platform based on neural network, *J. Phys.: Conf. Ser.*, **1650**, p. 032127, oct. 2020, doi: 10.1088/1742-6596/1650/3/032127.
[8] X Xu and M Duguleana, 2019 Trajectory Planning of 7-Degree-of-Freedom Manipulator Based on ROS, *IOP Conf. Ser.: Mater. Sci. Eng.*, **677**, p. 052072, doi: 10.1088/1757-899X/677/5/052072.
[9] I Kuhlemann, P Jauer, F Ernst and A Schweikard, 2016 Robots with Seven Degrees of Freedom: Is the Additional DoF Worth it?, *The 2nd International Conference on Control, Automation and Robotics*, 2016.
[10] A Vasquez, J Luis and Ordoñez Avila, 2020 Seven Degrees of Freedom Simulation Comparison for Industrial Process in San Pedro Sula, Honduras, *6th International Conference on Robotics and Artificial Intelligence*, **2020**, pp. 168-173. doi: 10.1145/3449301.3449330.
[11] P Corke, 2017 Robot Arm Kinematics, Robotics, Vision and Control: Fundamental Algorithms In MATLAB® Second, Completely Revised, Extended And Updated Edition, P. Corke, Ed. Cham: Springer International Publishing, 2017, pp. 193-228.
[12] W Liu, D Chen and J Steil, 2017 Analytical Inverse Kinematics Solver for Anthropomorphic 7-DOF Redundant Manipulators with Human-Like Configuration Constraints, *J Intell Robot Syst*, **86(1)**, pp. 63-79, abr. 2017, doi: 10.1007/s10846-016-0449-6

IOP Conf. Series: Materials Science and Engineering 1227 (2022) 012007 doi:10.1088/1757-899X/1227/1/012007

Development of an intelligent learning system based on fuzzy logic

V V Izvozchikova, T E Tlegenova and V V Markovin

Orenburg State University, Ave. Pobedy 13, Orenburg, Russia

E-mail: viza-8.11@mail.ru

Abstract. In modern institutions of higher education, the positive result of their work is measured by the progress of the student or by the factor how well the student meets the standard set by the state and the internal situation of the educational institution. In the current study, a new approach is proposed in assessing the progress of students, which is based on a fuzzy logic system. The proposed method uses three indicators: test tasks, written work and the possibility of mutual evaluation of students. Then, the quality of knowledge is assessed through the use of fuzzy logic, which is based on the Mamdani method. The results of applying this approach show that the proposed method can have practical application in assessing the quality of students' progress.

1. Introduction

The educational system, which is an actual field of activity, is determined by the high complexity of modeling the environment through mathematical expressions and the dynamics of variables, which greatly complicates the correlation process. The traditional form of interaction between a teacher and a student is still a verbal (informal) form. Over time, it will be possible to interpret human judgments in mathematical models, the description of which will take place in a natural, for a person, language, which will be based on fuzzy sets, with the help of which vague, divergent and incorrect criteria will be determined. The basis of the model of human actions and choices is well defined by fuzzy logic; this method is easily integrated into automated decision support systems. Using a mathematical model to build fuzziness in the initial information component, it will make it possible to obtain results that are adequate to reality.

In educational institutions, a positive result is determined by the student's academic performance or how well the student meets the established state educational standards. As the competition for careers becomes more intense every day, the importance of students doing well in all areas has attracted widespread public attention [1-8].

Educational assessment is a process in which learning outcomes are recorded, primarily in terms of measurable indicators of knowledge based on the use of certain criteria. Such an assessment can be defined as a measure of skills, attitudes and beliefs.

In order to effectively apply the fuzzy set in the field of education, many efforts have been made to determine an effective model. The purpose of the proposed method is to offer a new approach to the existing student assessment system.

2. Related work

To determine the relevance and effectiveness of the use of fuzzy logic in educational activities, a number of works were analyzed.

Troussasa et al. [9] developed and implemented an intelligent educational mobile application based on Quiz Time! to assess the level of knowledge of students based on a fuzzy logical model. The app includes a student assessment module, a vector-based personalized team game collaboration recommendation module, and a dynamic advice generator that uses fuzzy logic to help learners learn the C # programming language. The experiment involved 20 computer scientists and 80 undergraduate students of the Faculty of Informatics, the results confirmed the pedagogical adequacy of the application, its positive impact on learning and usefulness.

Under the guidance of Na Li [10], a recommendation model of fuzzy logic based on the fuzzy learning model (RGEL-FL) is proposed to assess the learning style, the level of the learning class and the degree of satisfaction from the training of engineering students. Data analysis based on the results of the study allowed us to build a logical model based on fuzzy logic and develop online training courses in order to activate the independent work of students. The approbation of the model allowed the authors to implement training, taking into account the individual preferences and capabilities of the student. The effectiveness, quality and accuracy of the training model are proved by taking into account the characteristics of a separate class of participants with the same preferences, their potential in a personalized educational environment. The proposed RGEL-FL method provides high efficiency.

In their study, F. Tugrul and M. Citil [11] applied the PROMETHEE multi-criteria decision making method (MCDM) using intuitionistic fuzzy set (IFS) to determine the ranking of school performance and select the most successful school depending on certain criteria. The students' GPA on the official exam was calculated and a grade point was awarded for each school. The official exam consisted of six core lessons. Thus, the criteria defined such basic lessons as Turkish, mathematics, science, social studies, English, religion. Simultaneous consideration of both intuitionistic fuzzy sets and criterion weights made it possible to obtain more consistent and rational results. The advantage of the PROMETHEE method lies in the assessment taking into account the criterion weight. The weight of the criteria shows how important they are. Through the use of the intuitionistic fuzzy method PROMETHEE, the success of the schools was assessed based on student achievement, and the most successful school was selected. The application can be used in cases where it is necessary to make a choice among alternatives based on a variety of criteria.

T. Aol et al. [12] developed an initiative-systematic assessment of the effectiveness (I-SPE) of leadership quality measurements based on the fuzzy logic approach. The study analyzed data within the framework of the leadership program - the Kepimpinan Pewaris Bangsa (PKPB) program conducted at the Universiti Teknologi MARA (UiTM) for highly rated students. The fuzzy logic apparatus was used to make decisions and evaluate the effectiveness of the approach in order to eliminate uncertainty and exclude the subjectivity of evaluating expert opinions. The authors proposed a model based on the production rules according to the criteria: leadership, communication, teamwork, discipline and the total average score for all criteria. The results obtained show that the proposed decision-making system makes it possible to fairly select the best candidate in comparison with the traditional method.

3. Methodology

A hybrid intelligent system has been designed to conduct research. This system is integrated into the teaching methodology of undergraduates of Orenburg State University and allows you to model the educational route of a particular student, taking into account his individual needs. For example, it is proposed to evaluate the overall level of mastering the discipline by means of the author's developed system based on the mathematical apparatus of fuzzy logic. For multidimensional models, decomposition operations are used, which allow dividing the model into many particular fragments of the model and conducting a complete analysis of the research area. The applied problem is solved by using the composition of the model on a fragment. In various applied subject areas that work in optimal or extreme situations, it may be necessary to make a decision in conditions of fuzziness, uncertainty, unreliability of time and resource constraints. The developed solution allows, based on the above factors, through the apparatus of methods and means of implementing dynamic interacting processes operating under conditions of uncertainty, to build a data management and processing system and offer an optimal solution.

Due to the high relevance of the parameter for evaluating the results of students, two types of control are proposed for the teacher: testing of students and conducting written control works.

Test indicators are achieved based on the objectivity of the measurement process and the interpretation of the results.

During the implementation of written control works, the teacher often treats subjectively or ambiguously to the subject of assessment. The implementation of the competence approach can provide an opportunity to solve such a problem. Such a decision implies a change in the way of assessment by means of a meaningful and criteria-based assessment of learning outcomes. A detailed study of the evaluation criteria (performance criteria) allows you to perform an unbiased procedure for evaluating learning indicators. The developed evaluation criteria are a list of the types of student's activities that he performs in the course of his educational activities, as well as optimal performance criteria. With such an implementation of the educational process, the subjectivism of the teacher is absolutely excluded or manifests itself to a small extent.

The values at the output of the variable "A.Test" (Testing of students) have the following input variables: a_1 – assessment of testing on topic 1; a_2 - assessment of testing on topic 2; ...; a_n – assessment of testing on topic n.

Let's define the indicators of the output variable "B.Teacher": in it, the teacher gives an assessment of the completed task according to the developed criteria for each topic: b_1 - assessment for the task on topic 1; b_2 - assessment for the task on topic 2; ...; b_n - assessment for the task on topic n.

In the output variable "C.Student": students themselves evaluate the work of their colleagues according to the proposed criteria: c_1 - evaluation of work on topic 1; c_2 - evaluation of work on topic 2; ...; c_n - evaluation of work on topic n.

In each variable ("A.Test", "B.Teacher" and "C.Student") determines the weight (significance) of the integral assessment. Component "B.Teacher" should have a higher significance index than other components, since the teacher is an expert in the subject area in terms of the discipline being taught. Therefore, it is advisable to set the following weight coefficients: "Test control" – 0,3; "Teacher" – 0,5; "Student" – 0,2.

The developed comprehensive solution for assessing the level of mastery of the discipline allows you to use a mathematical model associated with fuzzy sets and fuzzy logic. Such a solution is necessary when obtaining a result in conditions of limitation and fuzziness. A higher level of detail can lead to a reduction in the volume of processed and stored large amounts of data and to an increase in the performance of algorithms.

4. Mathematical model

According to Mamdani's algorithm, fuzzy logical inference is performed on a fuzzy knowledge base:

$$\bigcup_{p=1}^{k_j}\left(\bigcap_{i=1}^{n} x_i = a_{i,jp} \textit{with weight } w_{jp}\right) \rightarrow y = d_j, \quad (1)$$

in which the values of the input and output variables are given by fuzzy sets. We introduce the following notation necessary for further presentation of the material:

$\mu_{jp}(x_i)$ - function of the input belonging to a fuzzy set $a_{i,jp}$, i.e.:

$$a_{i.jp} = \frac{\int_{\underline{x_i}}^{\overline{x_i}} \mu_{jp}(x_i)}{x_i}; \quad (2)$$

$\mu_{dj}(y)$ - function of the output belonging to a fuzzy set d_j i.e.:

$$d_j = \frac{\int_{\underline{y}}^{\overline{y}} \mu d_j(y)}{y}. \quad (3)$$

The degree of membership of the input vector in which

$$x^* = (x_1^*, x_2^* \ldots, x_n^*) \quad (4)$$

belongs to fuzzy sets of d_j from the accumulated knowledge base, the calculation of this function is as follows:

$$\mu_{d_j}(X^*) = \vee_{p=\overline{1,k_j}} w_{jp} \cdot \wedge_{i=\overline{i,n}} [\mu_{jp}(X_i^*)]. \quad (5)$$

An operation from the s-norm (t-norm), i.e. from a set of implementations of logical operations OR (AND). Implementations are most often used: when performing the operation OR, the maximum is found, when performing the operation AND, the minimum is found.

As a result, we obtain such a fuzzy set y corresponding to the input vector x^*

$$y = \frac{\mu d_{1(X^*)}}{d_1} + \frac{\mu d_{2(X^*)}}{d_2} + \cdots + \frac{\mu d_{m(X^*)}}{d_m}. \quad (6)$$

The peculiarity of this fuzzy set is that the universal set for it is the set of the output variable y.

5. Practical part

Based on the results of a comprehensive assessment, as well as their interpretation (Table 1), a dynamic adjustment of the individual educational trajectory and the introduction of an intelligent system of recommendations is possible [13].

Table 1. Range of integrated assessment values.

The end result	Rating 'not satisfactory'	"Satisfactory" rating	Rating "good"	Rating "very good"	Rating "excellent"
Result	0-49	50-64	65-74	75-84	85-100

Let's consider the stages through which the results of mastering the discipline and making a comprehensive decision are evaluated under the condition of uncertainty in the developed fuzzy model.

1. Introduction of fuzziness on the initial range of values "A.Test", "Y.Instructor" and "C.Student", as well as the "output" value of "Result". With the introduction of fuzziness, the model transitions from a numerical indicator to a symbolic fuzzy value.
2. Writing rules by means of which the fuzzy inference algorithm will be applied using logical operations of union and intersection.
3. Representation of a fuzzy set of results obtained by a clear number in the value "Result".
   In figure 1 shows a scheme for evaluating the results of students, based on the use of a fuzzy model.

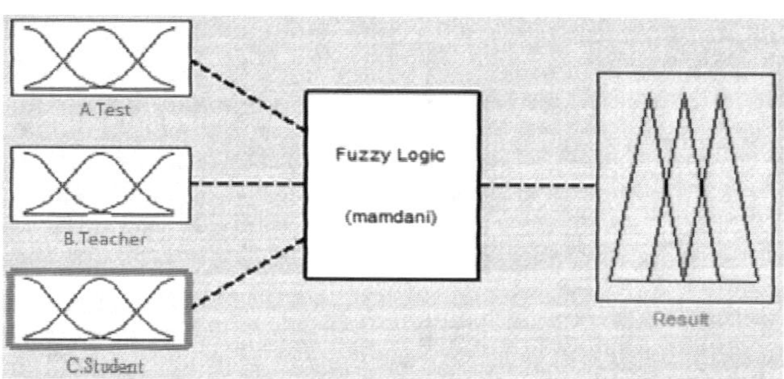

Figure 1. Scheme definition of evaluation of students' results.

Having conducted an experiment based on the developed course "Modern Tools", the structure of which is represented by five topics, it became possible to obtain a sets for each linguistic variable we are interested in.

When carrying out the developed testing method, it was decided to use a hundred-point scale for evaluating the results of mastering, the selected range of which is presented in table 2.

Table 2. A set of values in the variable "A.Test".

Final Result (Output)	Rating "Unsatisfactory"	Rating "Satisfactory"	Rating "Good"	Rating "Excellent"
A. Test	0-59	60-75	76-89	90-100

To facilitate the evaluation of test results for the teacher, certain evaluation criteria have been compiled for each group of tasks and in each topic.

Further, in accordance with the criteria obtained, scores were set and levels were identified (table 3). By analogy, students themselves evaluate the work of their colleagues.

Table 3. Values of sets "B.Teacher" and "C.Student».

Final Result	Rating "Unsatisfactory"	Rating "Satisfactory"	Rating "Good"	Rating “very good”	Rating "Excellent"
B. Teacher/C. Student	0-49	50-64	65-74	75-84	85-100

Rules are defined by membership functions, which are based on the results of input and output values. The developed rules are called linguistic. An example of the developed rule base is shown in figure 2.

1. If A.Test is Unsatisfactory and B.Teacher is Unsatisfactory and C.Student s is Unsatisfactory then Result is Poor
2. If A.Test is Unsatisfactory and B.Teacher is Unsatisfactory and C.Student is Satisfactory then Result is Poor
3. If A.Test is Unsatisfactory and B.Teacher is Unsatisfactory and C.Student is Very Good then Result is Average
4. If A.Test is Excellent and B.Teacher is Excellent and C.Student is Excellent then Result is Excellent
5. If A.Test is Good and B.Teacher is Very Good and C.Student is Very Good then Result is Very Good

Figure 2. Rule base.

The results of applying fuzzy logic using a rule base are shown in figure 3.

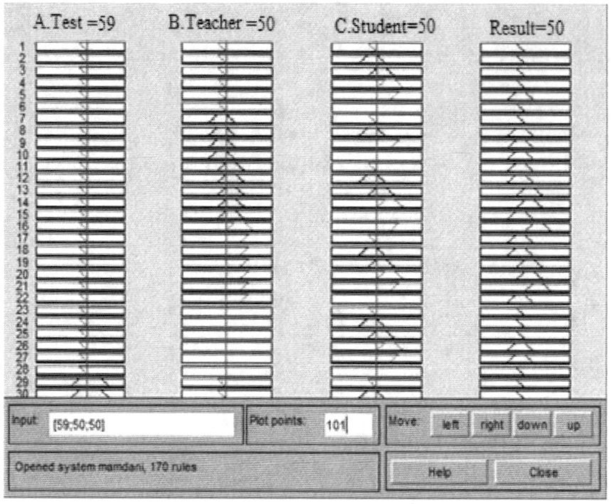

Figure 3. Results of applying fuzzy logic using a rule base.

6. Conclusion

In the presented work, a software package was used in the MATLAB environment, which contains functions for working with fuzzy logic, allowing you to build a logical inference model. In the work, three inputs were designed to set the initial range of values, which send a set of values through the processing module. The output of the results occurs at one output. The approbation of the developed model took place on different groups of students. The accuracy of the results is 95%. The developed model, based on the fuzzy logic method, allows:

- more accurately assess the level of training of students and the level of assimilation of the discipline;
- determine the most optimal input variables and set the most acceptable linguistic criteria;
- rebuild the developed model to meet different requirements and goals.

Further research involves the expansion of criteria and variables for more accurate analysis, as well as the introduction of this model into the system of online courses.

References

[1] Kurniawan D and Utama D 2021 Decision Support Model using FIM Sugeno for Assessing the Academic Performance *Advances in Science, Technology and Engineering Systems Journal* **6(1)** 605-11

[2] Panjaitan S and Alfajrin A 2021 Fuzzy Logic menentukan guru terbaik menggunakan metode Sugeno di Batam SMK Putra Jaya school *Computer and Science Industrial Engineering (COMASIE)* **5(6)** 69-77

[3] Nanda Jarti 2021 Decision Support System to determine students who are eligible to receive the scholarship of Indonesian Smart Card (KIP) By Using Fuzzy Sugeno *International Journal of Information System & Technology* **5(2)** 179-84

[4] Dam M, Majumder D, Bhattacharjee R and Santra S 2021 Performance measurement model for ranking of educational institutes: a fuzzy reasoning approach *Journal of Physics: Conference Series* **1797(1)** 012012

[5] Fointuna D W 2021 Applying Mamdani's method to categorize mathematical literacy of public middle school students in Kupang *Journal of Physics: Conference Serie* **1957(2021)** 012009

[6] Lima B N, Balducci P, Passos R P, Novelli C *et al.* 2021 Artificial intelligence based on fuzzy logic for the analysis of human movement in healthy people: a systematic review *Artificial*

Intelligence Review **54** 1507-23

[7] Ardhy F, Aminudin N, Zulkifli, Rizki F, *et al.* 2021 Implementation of Diabetes Mellitus Diagnosis Expert System Using Fuzzy Logic (Sugeno) Method Web-Based *International Journal of Grid and Distributed Computing* **14(1)** 270-81

[8] Tawarai F H and Fauziah 2021 Andrianingsih Web-Based Rice Disease Diagnosis Expert System Using Fuzzy Tsukamoto Method and K-Nearest Neighbor Algorithm *Journal of Computer Networks, Architecture and High Performance Computing* **3(2)** 153-60

[9] Troussasa C, Krouskaa A and Sgouropouloub C 2020 Collaboration and fuzzy-modeled personalization for mobile game based learning in higher education *Computers&Education* **144** 18

[10] Li Na, Chen X, Subramani S and Sifedine N 2020 Kadry Improved fuzzy-assisted multimedia *Assistive technology for engineering education* 12

[11] Tugrul F and Citil M 2021 A new perspective on evaluation system in education with intuitionistic fuzzy logic and PROMETHEE algorithm *Journal of Universal Mathematics* **4(1)** 13-24

[12] Ajol T, Gran S, Kanyan A and Lajim S 2020 An Enhanced Systematic Student Performance Evaluation Based on Fuzzy Logic Approach for Selection of Best Student Award *Asian Journal of University Education (AJUE)* **16(4)** 10-20

[13] Zaporozhko V, Shardakov V and Parfenov D 2020 Fuzzy model for evaluating the results of online learning *IOP Conference Series: Materials Science and Engineering* **734** 012150

Choosing technical components of the physical protection system of objects

S Yu Fedorov, V N Khalizev and E S Tarasov

Institute of Computer Systems and Information Security, Kuban State Technological University, Krasnodar, Russia

E-mail: fsu_50@mail.ru, ha53@mail.ru, helisar@mail.ru

Abstract. This article presents a method for solving the urgent problem of choosing a combination of technical systems of physical protection of objects of informatization of an organization, on the example of an integrated security system using methods of system analysis. The article proposes a new approach that combines the methods of the theory of antagonistic games and methods of fuzzy logic. On the basis of this mathematical model, a program and a database have been developed, including interconnected tables of physical protection systems, capabilities of protection systems, tables of threats and ways of implementing threats. As a result of the application of this method, with the help of the program, a specialist can quickly solve the complex problem of choosing the optimal combination of physical protection systems to counter probable threats for a specific object of the organization's informatization. Based on the results of the program's work, the developer receives a report with recommendations.

1. Introduction

The wide spread introduction of information systems and technologies provides new opportunities for solving social problems and opens up broad prospects for the development of the individual and society, but also aggravates old ones or generates new, previously unknown problems, becomes a source of new potential threats. At the moment, the acuteness of the problem of ensuring the information security of objects and protecting the information stored and processed in them from various threats is increasing. Without due attention to information security issues, the consequences of society's transition to new information technologies can be catastrophic. One of these problems is the provision of physical protection of information technologies and systems at the objects of informatization. Based on the analysis of international standards [1-4], it can be concluded that as one of the means of preventing information security incidents, it is necessary to use systems of physical protection of objects. Currently, there are a large number of such systems on the market, and the project developer has to make a personal decision to choose protection systems. The composition of physical protection systems can be varied depending on the threats, the purpose of the object, its significance and specific conditions. The problem is the lack of a unified regulatory framework and methods for choosing a means of physical protection of objects of the organization's informatization. The solution to this problem is a complex and urgent task due to the wide variety of such systems, the variety of protected objects, conditions and, accordingly, threats to them. The methodology proposed in this article allows to formally describing the process of choosing the means of physical protection systems for objects of informatization of an organization.

2. Materials and methods

Now almost every manufacturing enterprise, trade facility, government agency or office uses information technology and physical protection systems to ensure their safety. Traditionally, the set of physical protection systems included scattered components - video surveillance systems, security and fire alarms, and access control systems. Since about the mid-90s, consumer demands have become broader, and simple systems have ceased to solve more complex problems. Currently, in the Russian Federation, informatization objects are equipped with integrated security systems (ISS - hereinafter). An integrated security system is a developed specialized complex technical system that combines, on the basis of a single software and hardware complex with a common information environment and a single database, target functional technical subsystems and technical means designed for comprehensive protection of an object from standardized threats of various origins and manifestations, according to GOST R 53704-2009 [5] contains a list of 17 technical subsystems that should be included in the ISS: 1. duty dispatching office; 2. production and technological control; 3. security and alarm systems; 4. fire alarm; 5. control and management of access; 6. CCTV / video surveillance and control; 7. search subsystems; 8. fire automatics (fire extinguishing, smoke protection), 9. warning, evacuation); 10. connection with the object; 11. information protection; 12. physical barriers; 13. engineering support of the facility: electric lighting and power supply; 14. gas supply; 15. water supply; 16. sewerage; 17. maintaining the microclimate (heat supply, ventilation, air conditioning).

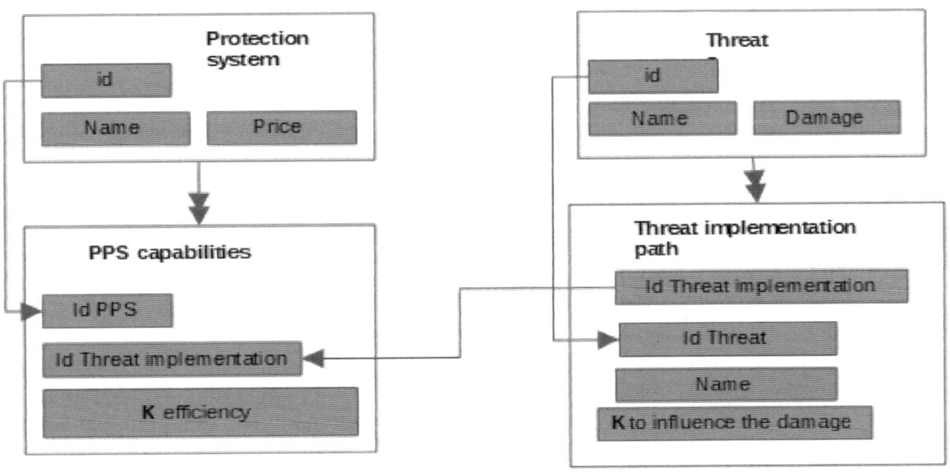

Figure 1. Database structure

The composition and number of these subsystems can vary. It depends on the purpose and importance of the protected object and the specific security conditions. The composition may include products from different manufacturers from different countries. One of the subsystems that can be included is the information security system. All system requirements must be taken into account by the developers when designing an integrated security system. The paper proposes to increase the selection accuracy to increase this number of subsystems to 22, by adding means to ensure the anti-terrorist security of an object, a mobile positioning subsystem (for mobile objects) and dividing the security alarm subsystem into a perimeter subsystem and a subsystem of object security equipment, as well as adding to the subsystems engineering support of the facility of backup power supply devices. Each subsystem has been assigned its own identifiers in the database table (figure 1). To solve the problem of choosing a set of subsystems, the ISS classification helps to a certain extent. ISS are subdivided depending on the required degree of automation of the facility and based on economic considerations. The first classification criterion is "the number of implemented basic protection functions. These ISS functions are exposed through its subsystems. So it can be: security alarm systems, fire alarm systems, alarm systems, fire extinguishing systems, access control and management, CCTV, etc. ISS must include at least two of the listed subsystems. The next criterion, the ISS classifications, are the principles of integration. The following types of subsystem integration are distinguished here: hardware, software and integration at the hardware-software level.

There is a huge variety of means of protecting computer information on the market, however, with an increase in the level of security of the system, certain inconveniences in its use, limitations and difficulties for users arise. Therefore, it is often necessary to choose the optimal protection option, which would not create great difficulties in using the computer. Based on the analysis of documents, it can be concluded that the process of analysis and synthesis of integrated security systems, the choice of their subsystems is rather complicated and time-consuming, depending on a considerable number of various factors, the amount of knowledge and experience of designers. Therefore, errors may appear due to various reasons, which can lead to the choice of a deliberately expensive or not the best option, which can then affect the budget and safety of the entire facility. Therefore, the task of this formalized choice of the optimal composition of the set of integrated security systems (ISS) is relevant at this stage.

In this method, at the first stage, it is proposed to use the mathematical apparatus of the theory of matrix antagonistic games, which makes it possible to optimize the choice of means of physical protection of objects of informatization of an organization. System selection methods are divided into universal and special. As a rule, universal ones are less accurate, they do not reflect the specific features of the subject area, etc., therefore it is better to use special ones. It is proposed to use the methods of game theory, because in reality and in life there are two opposing sides - an attacker from whom threats emanate and an information security specialist who uses various means of protection to counter threats, an intruder. We used the example program described in the article [6]. The goal of the offender is to inflict maximum damage on the attacked computer system, and the goal of the information security specialist is to select a protection strategy in which the probable losses from information security incidents will be minimal. That is, the physical protection system must be ready to neutralize information security incidents that cause maximum damage on the part of the intruder. As a payment matrix, a game table was disassembled, in which the columns are the strategies of the information security specialist and are filled with the used physical protection means y_j (where j = 1, m), and the rows represent the strategies of the violators and the threats x_i are filled (where i = 1, n) of some matrix. As a strategy of the violator, various threats to physical protection objects presented in [5] were taken as a basis. The result of the game aij is the total damage that can be caused to the computer system when the violator implements his i-th strategy, as well as the damage inflicted to the IS specialist from the cost of the total cost of protective equipment from the j-th strategy. The desired value is the sum of the possible damage from the remaining attacks from the attacker's strategy and the cost of physical protection equipment. Given the available information on information security incidents, the result will be a strategy for physical protection systems that minimizes average losses, i.e. the minimum amount will be:

$$a_{i0j0} = \min_{i} \max_{j} a_{ij} \quad (1)$$

All possible combinations of the following 21 threats were selected as possible strategies of the violator in this work. It was possible to obtain various strategies of the players by a simple enumeration of all possible information security incidents for an intruder and a similar enumeration of all physical protection software tools participating in the game for an information security specialist. To reduce the number of players' strategies, threats and physical protection were divided into groups. In this case, from each group of threats, one with the maximum damage is selected, and from each group of protection tools, one with the minimum cost is selected. From the resulting lists of incidents of information security and protection means, the strategies of the players are drawn up by enumerating all possible combinations. The total costs are obtained by summing the amount of damage that can be caused when implementing the current strategy of the attacker, if the system was not protected from it by means of protection from the current strategy of the information security specialist, and the total cost of all means of protection from the current strategy of the security specialist. The calculation of damage from the implementation of threats is calculated in two stages. First, the attacker's current strategy is matched against each of the physical defenses from the security specialist's current strategy, and if the agent protects against any threats from the current set, then these threats are removed from the set. By comparing the current set of threats with all the means from the current strategy of the information security specialist, a certain number of threats are obtained, from which the system is not protected in this case. The received threats must be compared with all the existing threats and the damage values of those threats that are present in the received set must be summed up. Further, these two sums are added, and damage is obtained when the current pair of strategies of the violator and the information security specialist is applied [6]. The algorithm presented in [6] was refined using fuzzy logic and allows you to find the most dangerous ways to implement threats and select physical protection systems (PPS) based on a given list of intruder threats for a specific object and a specific situation. Rows have been added to the threat table in the database, containing the paths of the threat implementation. For example, an incidentsuch as theft can be implemented in 2 ways:

Option 1 - by hacking (using special tools):

- through the balcony (id 10);
- through the basement (id 11);
- through the window (id 12);
- through the roof (id 13);
- through the door (id 14).

Option 2 - no hacking:

- through an unlocked front door (id 15);
- through an unlocked window id (16).

At the first stage, the maximum is determined. For each threat in the PPS combination, the value of non-avoidable damage is determined, which is initially assumed to be equal to the maximum possible damage caused without the use of PPS and decreases in the process of considering the capabilities of the PPS in combination to counteract their threats. To do this, first, the value of d is found - the unavoidable maximum damage caused by a specific way of implementing the threat according to the formula (2):

$$d = U * k, \quad (2)$$

where d - is the unavoidable maximum damage caused by a specific way of implementing the threat;

U - is the maximum damage to the threat realization; k - is the coefficient of effectiveness of the implementation of this threat along the given path (where $k = 0 \dots 1$).

For each PPS in a possible combination

$$NU = d - (d * kp), \quad (3)$$

where NU - is the non-preventable damage for this threat.

kp - coefficient of PPS counteraction to the given method of realization of the given threat.

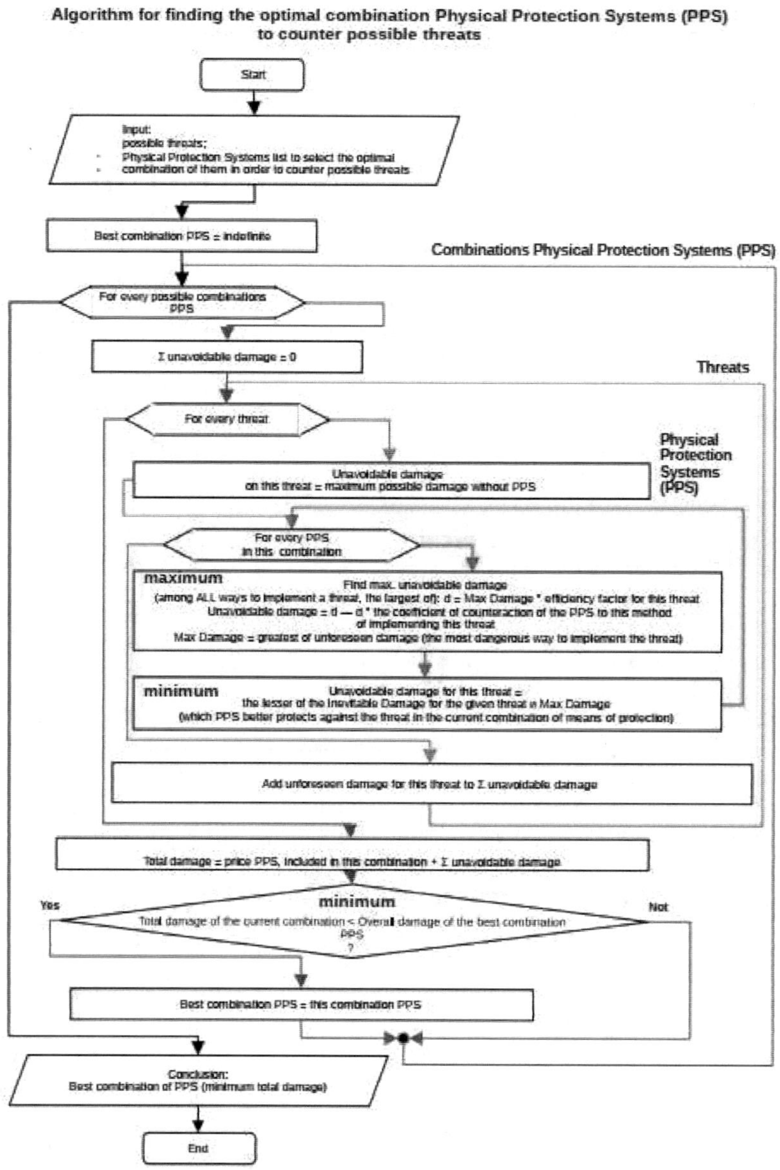

Figure 2. Algorithm for finding the optimal combination of physical protection to counter possible threats Algorithm for finding the optimal combination of physical protection to counter possible threats

Non-preventable damage for a given threat is determined by formula (3), as the difference between the maximum damage that cannot be prevented by a specific way of implementing the threat and the prevented damage, which is calculated as the product: *(d * kp)*. For each threat, by comparing all values, the greatest value is found, this will be the most dangerous way of implementing the threat, that is, the way through which the greatest damage is possible. At the next stage, the minimum is found. By enumerating and comparing all PPS values, we find the best PPS combination that provides the best security for a given threat in the current combination. Among all physical protection systems, there is the PPS that has the lowest value of the combination of non-avoidable damage covered by the implementation of a given threat and the maximum damage that can be caused by a threat to a given physical protection. Next, the sum of the minimum non-preventable damages for all threats is calculated. Then the total damage for the PPS combination is determined by summing up the non-preventable damage for all threats and the cost of all PPS included in this combination. To obtain the best combination of PPS, which ensures the greatest security of the informatization object, the combination with the lowest value of the total damage is selected by enumerating and comparing the current obtained combination of PPS with the total damage of the already existing best combination of PPS.

At the last step, the choice of physical protection subsystems to counter possible threats is carried out based on the application of the concept of building a protection system in the form of covering a set of functional requirements and developing a complex preference indicator according to a modified target method.

Table 1. Sample report. The optimal list of protection tools to prevent existing threats Results and discussion

Means of protection	Tele / video surveillance and video monitoring subsystem	Alarm subsystem
Cost	40 000.00	205 943.00
Threat 1	Theft	
Path of threat realization	Through the roof	Through an open window
Unavoidable damage	60 000.00	360 000.00
Threat 2	Robbery	
Path of threat realization	Without violence against the person	With violence against the person, not dangerous to life and health
Unavoidable damage	400 000.00	270 000.00
Threat 3	Sabotage	
Path of threat realization	economic	cyber diversion
Unavoidable damage	270 000.00	180 000.00
Threat 4	Terrorism	
Path of threat realization	Organized, collective	Organized, collective
Unavoidable damage	1 350 000.00	270 000.00

The total cost of protective equipment: 245 943.00. Total uncovered damage from threats: 780,000.00. The maximum costs will be no more than: 1,025,943.00.

A sample report is presented in table 1, which shows the optimal combination of physical protection means to counter potential threats for a specific object of the organization's informatization and the total costs of the selected systems. In addition, the report calculates the maximum costs for a complete set of all protection measures, which significantly exceed the costs of the combination selected by the program. The program also provides information on the most dangerous ways of implementing threats.

3. Results

Based on the described approach using the mathematical apparatus of antagonistic game theory and fuzzy logic, the following results were obtained:

1. An algorithm (figure 2), a database, a software application in the C ++ programming language have been developed using a cross-platform integrated development environment IDE Qt Creator version 5.12.0.

2. Compiled a list of 17 threats and 21 physical protection subsystems.
3. The correctness of the developed mathematical model was checked.
4. The choice of the optimal combination of physical protection systems was made to ensure the security of a specific object of the organization's informatization and to counter potential threats in the event of information security incidents.
5. The most dangerous ways of implementing threats have been determined, based on the analysis of a potential list of threats for a given object.
6. Determined the maximum costs for a full set of physical protection of objects of informatization of the organization and the cost of an optimal set of physical protection. At the same time, significantly reducing the total costs.

4. Conclusions

Thus, the developed mathematical model and the program written on its basis allow solving the urgent problem of choosing the optimal set of physical protection means based on the analysis of a large amount of information (17 threats and 21 physical protection subsystems) to counter potential threats for a specific object of the organization's informatization.

References

[1] ISO / IEC TR 18044: 2004 Information technology - Security techniques - Information security incident management".
[2] ISO / IEC 27035-1: 2016 - Information technology - Security techniques - Information security incident management - Part 1: Principles of incident management.
[3] ISO / IEC 27035-2: 2016 - Information technology - Security techniques - Information security incident management - Part 2: Guidelines to plan and prepare for incident response.
[4] ISO / IEC 27035-3: 2020 Information technology - Information security incident management – Part 3: Guidelines for ICT incident response operations.
[5] GOST R 53704-2009: Complex and integrated security systems. General technical requirements.
[6] Vakhniy T V and Guts A K, Novikov N Yu. 2016. Matrix-game program with the choice of criterion for determining the optimal set of security tools for a computer system. Omsk State University. Dostoevsky. *Mathematical structures and modeling.* **2 (38).** 103-15
[7] Garcia M 2002. *Design and assessment of physical protection systems.* (M.: Mir: LLC "AST Publishing House").

IOP Conf. Series: Materials Science and Engineering 1227 (2022) 012009 doi:10.1088/1757-899X/1227/1/012009

Development of a bi-fuel power supply system for a diesel engine

Sh V Buzikov [1*], O P Matushkin [2] and Yu N Matushkina[3]

[1] Department of Construction Production, Vyatka State University, Kirov, st. Moskovskaya, 36, Russia
[2] Department of Materials Science and Design Basics, Vyatka State University, Kirov, st. Moskovskaya, 36, Russia
[3] Department of Economics, Vyatka State University, Kirov, st. Moskovskaya, 36, Russia

*E-mail: shamilvb@mail.ru

Abstract. The relevance of the study is due to the lack of systems in which the proportions of the used fuels in the mixture and the values of the cycle supply are simultaneously controlled depending on the load-speed mode of the diesel engine. The aim of the research is to develop a design for a bi-fuel power supply system for a diesel engine, which allows for the simultaneous regulation of the proportions of the components of a mixture consisting of alternative and diesel fuels, as well as the values of the cyclic supply of this mixture to the diesel cylinders. As a result of the research, an algorithm was calculated for the simultaneous control of the proportions of the components of a mixture consisting of alternative and diesel fuels, as well as the values of the cyclic supply of this mixture to the diesel cylinders. As a result of the operation of the fuel supply system on mixtures of diesel and alternative fuels according to the developed algorithm, it made it possible to achieve the identical operation of a diesel engine on pure diesel fuel at all load-speed modes of diesel operation. Also, in the course of the research carried out, a bi-fuel power supply system for a diesel engine was developed, which makes it possible to simultaneously regulate the amount (cycle supply) and the composition of a mixture consisting of diesel and alternative fuel supplied to the diesel cylinders, depending on the speed and load modes of operation.

1. Introduction

Alternative fuels used in diesel engines today occupy a special place among other renewable energy sources around the world [1]. They owe their attention to the stable long-term positive global price dynamics of traditional diesel fuel and the annually stricter environmental requirements for diesel engines [2].

Among the named alternative fuels, liquid alternative fuels such as benzene, alcohol, oils, ethers, glycerin, etc. have become quite widespread. One of the main and indisputable advantages of using alternative fuels in comparison with diesel fuel is their environmental friendliness, which provides an incentive for the further development of diesel engines in the era of "green technologies" [2]. Along with high environmental friendliness, the use of such fuels is associated with the deterioration of the effective performance of diesel engines [1, 2]. Their use in their pure form is difficult due to the difference in their properties compared to diesel fuel [2].

Today, the method of using liquid alternative fuels in conjunction with diesel fuel is gaining more and more scientific interest [2]. The most common way here is the joint use of both alternative fuels and traditional diesel fuel [2].

In studies [3-6], the supply of alcohol-containing and vegetable fuel was carried out together with diesel fuel by pre-mixing them in certain proportions until a homogeneous mixture was obtained, followed by refueling of the diesel engine.

In works [7-11], studies were carried out on the use of various alternative fuels in diesel engines together with diesel fuel simultaneously in different proportions and different load-speed operating conditions.

In the studies carried out [12-14], diesel fuel supply control systems were developed and applied, in which the cyclic fuel supply was regulated depending on the proportion of fuels used in the mixture and the load-speed operating mode.

However, at this point in time, there are no systems in which the proportions of the used fuels in the mixture and the values of the cycle supply are simultaneously controlled depending on the load-speed mode of operation. Therefore, the development of a design for a bi-fuel power supply system for a diesel engine, which will allow the above-described regulation to be carried out, is very relevant.

2. Research objectives

The purpose of our research is to develop a design of a bi-fuel power supply system for a diesel engine, which allows simultaneous control of the proportions of the components of a mixture consisting of alternative and diesel fuels, as well as the values of the cyclic supply of this mixture to the cylinders. To achieve this goal, it is necessary to solve the following tasks. First, to develop an algorithm for the simultaneous control of the proportions of the components of a mixture consisting of alternative and diesel fuels, as well as the values of the cyclic supply of this mixture to the diesel cylinders. Secondly, to develop a design of a bi-fuel power supply system for a diesel engine to implement the above-mentioned operation algorithm.

3. Materials and methods

Normal operation of a diesel engine on various mixtures of alternative and diesel fuel seems to be quite problematic without introducing a number of design changes to the fuel supply control system.

After analyzing the characteristics of the fuel supply shown in figure 1, it can be seen that with the fuel supply equipment installed by the manufacturer, the dependence of the heat Q introduced into the diesel engine with clean diesel fuel on the stroke of the high-pressure fuel pump rack h is represented by line 1.

When the fuel supply equipment operates on various mixtures of diesel and alternative fuels, the characteristic of the high-pressure fuel pump changes in comparison with the operation on pure diesel fuel. Taking into account that the specific lower design heat of combustion of a mixture consisting of diesel and alternative fuels is less than that of pure diesel fuel, the dependence of heat input into diesel cylinders on the rail travel is represented by line 2 in figure 1.

After analyzing these two dependences of lines 1 and 2 shown in figure 1, it can be seen that at the same position of the rail of the high-pressure fuel pump, the heat introduced into a diesel with a fuel consisting of a mixture of diesel and alternative fuels is less compared to pure diesel fuel. Consequently, when the rail of the fuel pump of weighty pressure is in position, taking into account the action of the regulator of the corresponding nominal flow and the overload mode, the required amount of heat with fuel set by the manufacturer of the fuel supply equipment will not be introduced into the diesel engine.

This circumstance will lead to a decrease in the nominal torque safety factor and a decrease in the ability to overcome the overload, as well as a violation of the high-speed operation of the diesel engine.

IOP Conf. Series: Materials Science and Engineering 1227 (2022) 012009 doi:10.1088/1757-899X/1227/1/012009

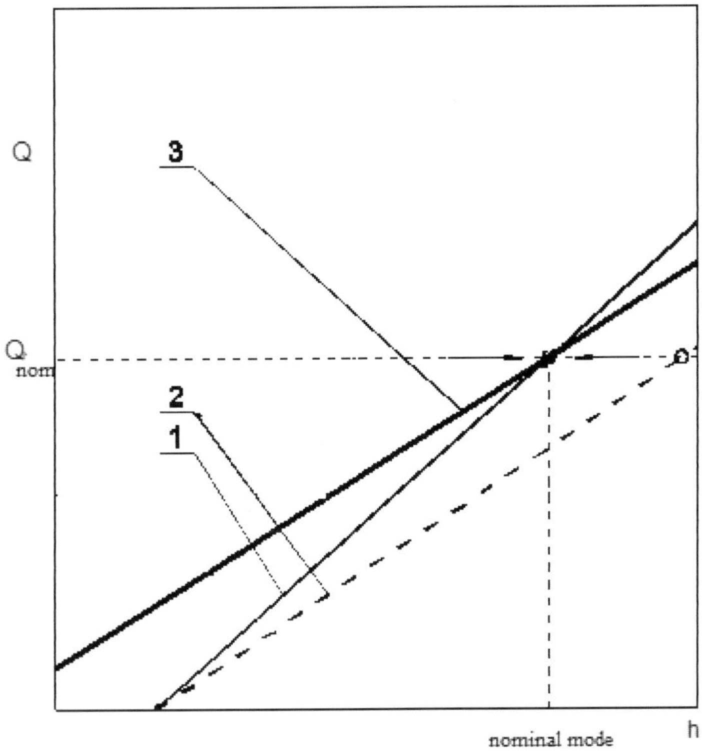

Figure 1. Fuel supply characteristics: 1 - dependence Q = f (h) for diesel fuel; 2 - dependences Q = f (h) for a mixture of diesel and alternative fuels; 3 - dependences Q = f (h) for a mixture of diesel and alternative fuels, taking into account changes in the design of the high-pressure fuel pump.

Elimination of this circumstance is possible by making changes to the design of the high-pressure fuel pump, the dependence of which is determined by line 3 in figure 1. Analysis of this dependence shows that in the nominal operating mode the amount of heat introduced into the diesel with fuel corresponds to the fuel supply equipment installed by the manufacturer. However, in the overload mode, there is also a decrease in the nominal torque safety factor and a violation of the high-speed operation of the diesel engine in the partial load mode, as well as the impossibility of turning off the supply. The moment when the regulator starts to operate, set by the manufacturer of the fuel supply equipment, will also not allow the diesel to be brought to the nominal operating mode.

Therefore, it is necessary to make changes in the design of the high pressure fuel pump, which will lead to the operation of the fuel supply system according to the following algorithm. With the position of the high pressure fuel pump rack corresponding to the rated operating mode of the diesel engine and overloading, clean diesel fuel will be supplied to ensure the performance specified by the manufacturer. In the range of the rack positions from the minimum feed to the nominal feed, the mixture of fuel, consisting of diesel and alternative, will change proportionally. That is, the greater the stroke of the rack h, the greater the content by weight of diesel fuel in the mixture. With the minimum rack stroke h in the mixture, the content by mass of the alternative fuel will be maximum. With a subsequent increase in the stroke of the rack h up to the nominal one, the content of diesel fuel in the mixture increases up to 100%.

As a result, the operation of the fuel supply system on mixtures of diesel and alternative fuels in compliance with the calculated ratios will make it possible to achieve the identical operation of a diesel engine on pure diesel fuel at all load-speed modes of operation.

4. Results and discussion

A significant difference between the proposed design of the diesel power system from all previously known solutions is that a groove is made on the forked lever 7 for the possibility of free movement along it of the hinge end of the intermediate link 8, which is connected with the hinged end of the main lever 3 with the help of an additional spring 21, moreover, the drive lever 19 through the two-arm lever 20 is connected to the articulated end of the main lever 3 in figure 2.

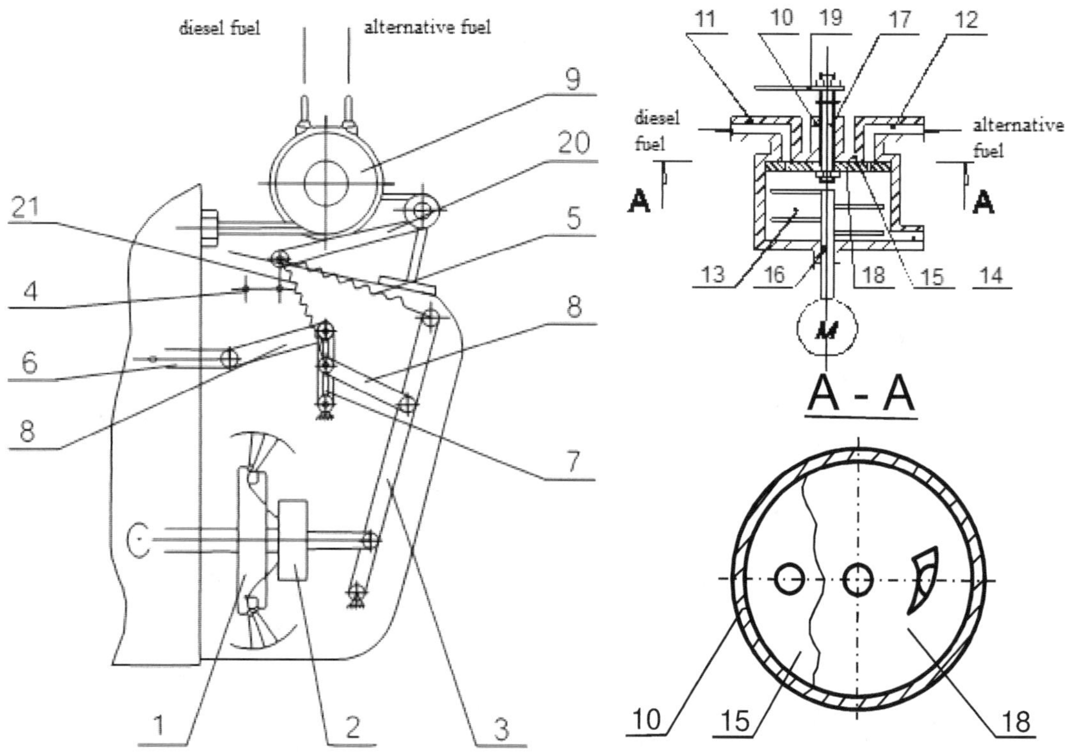

Figure 2. Bi-fuel power supply system of a diesel engine: 1 - centrifugal sensor; 2 - movable coupling; 3 - the main lever; 4 - control lever; 5 - spring; 6 - dosing organ; 7 - forked lever with a groove; 8 - intermediate rods; 9 - mixer-dispenser of fuels; 10 - case; 11 - channel for supplying diesel fuel; 12 - channel for supplying alternative fuel, 13 - mixing chamber; 14 - channel for removing a mixture of fuels; 15 - wall; 16 - mixer; 17 - shaft; 18 - disk; 19 - drive lever; 20 - two-armed lever; 21 - additional spring.

The control system of a multi-fuel diesel engine works as follows in figure 2. Depending on the speed and load mode of the diesel engine and the position of the control lever 4, the centrifugal sensor 1 with the movable clutch 2 sets the main lever 3 to the position of the dynamic balance of the spring force 5 and the force of the centrifugal sensor 1. Main lever 3 through a forked lever with a groove 7 and an intermediate rod 8, depending on the position of the control lever 4 with the help of an additional spring 21, dynamically set the kinematic relationship between the intermediate rods 8, depending on the position of the control lever, thereby providing automatic correction of the amount of mixed fuel supply by the metering element 6 In this case, the control lever 4 through the hinged end with the help of the two-armed lever 20 moves the drive lever 19 in the mixer-dispenser 9, forcing the working shaft 17 with the disc 18 with slots attached to it to rotate. The disc 18 with slots, turning, changes the flow cross-sections of the holes in the wall 15 of the housing, thereby changing the volumetric amount of diesel and alternative fuel entering the mixer-dispenser through channels 11 and 12. As a result, the movement of the main lever 3, depending on the position of the control lever 4, is simultaneous regulation of the quantity and composition of the mixture at all high-speed and load modes of operation of the diesel engine.

The operation of such a system will make it possible to simultaneously regulate the amount (cycle supply) and the composition of the mixture, consisting of diesel and alternative fuel, supplied to the diesel cylinders, depending on the speed and load modes of operation.

5. Conclusion

As a result of the research, the algorithm of operation of the fuel supply equipment, which is part of the general diesel power system, was calculated, which allows you to simultaneously adjust the proportions of the components of the mixture consisting of alternative and diesel fuel, as well as the values of the cyclic supply of this mixture to the diesel cylinders. The design of a bi-fuel diesel power supply system for the implementation of the above-mentioned operation algorithm has been developed.

References

[1] Wagner V A 2000 The use of alternative fuels in the internal combustion engine *Bulletin of the Altai State Technical University named after I.I. Polzunov* **2** 77-86.

[2] Likhanov V A 1998 The use of alternative fuels in tractor diesel engines *Concept of development of mechanization, electrification and automation of the agro-industrial complex of the North-East: Sat. materials scientific and practical. conf* (Kirov) pp. 48-50.

[3] Rahiman M K, Santhoshkumar S, Subramaniam D, Avinash A and Puga-zhendhi A 2021 Effects of oxygenated fuel pertaining to fuel analysis on diesel engine combustion and emission characteristics *Energy* **239** D. doi.org/10.1016/j.energy.2021.122373.

[4] Agarwal A K and Chandra K 2021 Di-ethyl ether-diesel blends fuelled off-road tractor engine: Part-I: Technical feasibility *Fuel* **308**. doi.org/10.1016/j.fuel.2021.121972.

[5] Asad U, Divekar P S and Zheng M 2021 High efficiency ethanol–diesel dual-fuel combustion: Analyses of performance, emissions and thermal efficiency over the engine load range *Fuel*. doi.org/10.1016/j.fuel.2021.122397.

[6] Sittichompoo S, Theinnoi K, Sawatmongkhon B, Wongchang T, Iamcheerangkoon T and Phugot S 2021 Promotion effect of hydrogen addition in selective catalytic reduction of nitrogen oxide emissions from diesel engines fuelled with diesel-biodiesel-ethanol blends *Alexandria Engineering Journal*. doi.org/10.1016/j.aej.2021.11.002.

[7] Duan X, Xu Z, Sun X, Deng B and Liu J 2021 Effects of injec-tion timing and EGR on combustion and emissions characteristics of the diesel engine fuelled with acetone–butanol–ethanol/diesel blend fuels *Energy* **231** doi.org/10.1016/j.energy.2021.121069.

[8] Mansor M R A, Mohamad T I and Sabah O 2021 Numerical investiga-tion on combustion and emissions in a direct injection compression ignition engine fuelled with vari-ous hydrogen–methane–diesel blends at different intake air temperatures *Energy Reports* **7** 403-21 doi.org/10.1016/j.egyr.2021.07.110.

[9] Agarwal A K, Valera H and Mustafi N N 2021 Di-ethyl ether-diesel blends fuelled off-road tractor engine: Part-II: Unregulated and particulate emission characteristics *Fuel* **308** doi.org/10.1016/j.fuel.2021.121973.

[10] Dhahad H A, Fayad M A, Chaichan M T, Jaber A A and Megaritis T 2021 Influence of fuel injection timing strategies on performance, combustion, emissions and particulate matter characteristics fueled with rapeseed methyl ester in modern diesel engine *Fuel* **306** doi.org/10.1016/j.fuel.2021.121589.

[11] Attia A M A, KulchitskiyA R, Nour M, El-Seesy A I, Nada S A 2021 The influence of castor biodiesel blending ratio on engine performance including the determined diesel particulate matters composition *Energy* **239** A. doi.org/10.1016/j.energy.2021.121951.

[12] Bae G, Choi S, Lee S, Moon S, Wang J, Kim K, Lee J and Kang J 2021 Experimental investigation of fuel temperature effects on transient needle motion and injection velocity of

solenoid type diesel injector *International Journal of Heat and Mass Transfer* **181** doi.org/10.1016/j.ijheatmasstransfer.2021.121838.

[13] Yang M, Ren C, Jiang S, Xin Y, Hu Y and Liu Z 2021 Theoretical predictions of compatibility of polyoxymethylene dimethyl ethers with diesel fuels and diesel additives *Fuel*, **307** doi.org/10.1016/j.fuel.2021.121797.

[14] Liu D, Li X, Shang H, Xie L, Chen Y and Chang J 2021 Combus-tion performance and fuel injection timing adaptability of a lateral swirl combustion system for di-rect injection diesel engines *Fuel* **296**. doi.org/10.1016/j.fuel.2021.120663.

[15] Patent No. 2651021 C1 Russian Federation, IPC F02M 43/00, F02M 31/18, F02M 27/08. Diesel fuel supply system with additional fuel: No. 2017120445: app. 06/09/2017: publ. 04/18/2018.

Analytical studies of plant proteins based on grain mixtures

I Kechkin[1], V Ermolaev[2], M Yakovchenko [2], E Gurkovskaya[3] and E Makhacheva[4]

[1] Plekhanov Russian University of Economics, Academic Department of Catering Business Stremyanny lane, 36, Moscow, 117997, Russian Federation
[2] Kuzbass State Agricultural Academy, 5, Markovtsev Str., Kemerovo, 650056, Russia
[3] K.G. Razumovsky Moscow State University of technologies and management (the First Cossack University), 73, Zemlyanoy Val str., 109004, Moscow, Russia
[4] Department of Technology and Organization of Food Production, Faculty of Business, Novosibirsk State Technical University (NSTU), 20, K. Marksa Ave., 630073 Novosibirsk, Russia

E-mail: kechkin87@mail.ru, ermolaevvla@rambler.ru, mara.2002@mail.ru, e.a.gurkovskaya@yandex.ru

Abstract. Based on the result analysis obtained by grinding the grain mixture on coarse systems of the break process, a technological scheme of grinding was developed, it includes three break systems, a scratch system, a grinding system and five reduction systems, which ensured a flour yield of 70 - 75%.After developing the technology of the grinding process for wheat grain with the aim to obtain an optimal flour yield of 70-75%, and the preparation of binary mixtures of composite types of flour with specified technological properties, there was analyzed the samples' quality in terms of molecular weight (MM) to confirm their nutritional value. The analysis was carried out to determine the protein molecular mass of wheat grain, flax seeds and wheat flour on the basis of the developed TSEC technique using polyethylene glycols (PEGs) as standards and marker proteins to identify the analysis results. Detailed analytical studies of plant proteins for obtaining qualitative and quantitative characteristics by the TSEC method are presented in the paper. Analytical characteristics of the method have been studied, which makes it possible to obtain reproducible results of a kind of qualitative analysis of biopolymers by quantitative characteristics of molecular weights and quantitative analysis of proteins by optical densitometry. The developed methods for determining the molecular mass of proteins are characterized from the metrological point of view.

1. Introduction

Today food quality is considered as a factor of national security, therefore, the determination of the raw material chemical composition, structure and supramolecular structure is an urgent task for the commodity assessment of plant proteins in food production technologies. In the diet of the population of our country, bakery products, most of which are products based on various types of flour and cereals, occupy a significant share (up to 40%). In this regard, it seems very promising and in demand to form flour composite mixtures at flourmills with a given content of basic nutrient and biologically active substances: proteins, carbohydrates, fats, vitamins, micro- and macro-elements from grain processing products, incl. wheat, triticale, rye, barley and chickpea. Such mixtures will become the basis for specific manufacturers producing bakery, confectionery, pasta and extruded products, dairy and meat products of functional nutrition: dietary, prophylactic and therapeutic purposes [1].

Determination of the molecular weight distribution (MWD) of native polymers and oligomers is a decisive moment in the science of biological properties of cereals. A special place is occupied by the assessment of the content and complexation of proteins, since proteins are the most important factor in the nutritional value of food products.

Based on the use data obtained with the latest research methods for high-polymer compounds in grain science, the species and varietal specificity of the protein components of plant raw materials and their role in the process of processing into final food products are shown. The nutritional value of grain products depends on the composition and processing technology. According to the chemical composition and properties of the individual components, the grains of cereals of different tribes (tribes) differ from each other; the differences in protein substances are especially noticeable, the properties of which characterize representatives of individual tribes no less than their morphological characters [2].

The main analytical parameters of proteins in food are their molecular weight, amino acid composition, structure and quantitative ratio of various fractions. One of the most powerful methods for the isolation and determination of protein fractions is the method of Thin-layer Size Exclusion Chromatography (TSEC), which is currently the most effective method for studying proteins, since it differs in a number of necessary features: independence from the pH of the medium, temperature, preservation of the native composition, properties and pattern structures.

The purpose of this study is to prepare binary grain mixtures (wheat + flax) to obtain composite types of flour with given technological properties, as well as to develop a technique for analytical work of the TSHC method for analyzing plant proteins in terms of molecular weight (MW) and determining their qualitative characteristics. [3]

2. Materials and methods

Analysis of some physical properties, in particular, geometric dimensions of flax seeds and wheat grains, shows.

To determine the optimal values of the conditioning parameters, the method of contour-graphic analysis was used. The following optimization criteria were used:

- estimated output of premium flour.
- the maximum value of flour whiteness.
- the ratio of the bran yield of break systems to the bran yield of reduction systems, as a characteristic of grinding system efficiency.

The independent variables in the equation are grain moisture and dwell time shown in Table 1.

Based on the results of the study, it can be concluded that the introduction of flax seeds into the mixture does not have a significant effect on the choice of conditioning parameters. The technology for preparing the grain mixture "*wheat + flax*" is possible immediately before grinding, which is due to the separate preparation of the components and the self-sorting of the mixture while movement [4,5].

Preparation of a grain mixture for mixing and formation of a homogeneous composition during its processing, taking into account significant differences in physical properties of the components, is a rather difficult task. The stability of such a mixture depends on similarity of the component physical properties, first of all, the size and shape of the particles, which are fundamental characteristics of bulky material.

For this reason, experimental experiments were carried out to determine the necessary formation conditions for the binary grain mixture and then there were modeled the main mixing methods:

- active methods - with a high relative speed of component movement, which are based on a convective movement mechanism (paddle mixers),
- passive ones - based on the movement of layers' sliding relative to each other (drum mixers).

The obtained quality of mixtures was assessed by the coefficient of heterogeneity (coefficient of variation).

$$V = \frac{100}{\bar{c}} \sqrt{\frac{\sum_1^n (c_i - c)^2}{n-1}} \quad (1)$$

where $\bar{c}$ - arithmetic mean of the key component concentration,
c_i- current concentration value,
n – number of measurements.

The samples' number and weight were determined in accordance with the recommendations [1] and amounted to: number - 8; weight by calculation -15g, actual weight - 50g.

Comparison of the methods of grain mass mixing showed that the passive method is significantly inferior to the active one; therefore, the active method of mixing was used, which ensured a satisfactory quality level of the product obtained.

3. Results

Analysis of the grinding process and croup formation. Grinding modes. At the first stage of the study, the approximate values of the modes of grain mixture grinding for the 1-3 break systems (I-III br.s.) were determined.

Based on the calibration charts, the values of the roll mill gaps have been established, which provide the extraction range from 25-70% for the I break system and 45-70% for the II break system. The roll gap varied from 0.75mm to 0.20mm on the I break system, from 0.2mm to 0.05mm on the II break system and from 0.05mm to 0.00 on the III break system.

The main part of the intermediate grinding products of the grain mixture lies in the particle size range from 600 to 150 μm, which later determined the set of sieves for the analysis of gross products, the formation of the milling technological scheme and the production of the final product.

Optimum modes were achieved with the following roll mill gaps: I break system - 0.40mm; II break system - 0.20mm; III break system -0.05mm.

Analysis of the grain mixture quality

After developing the technology of the grinding process for wheat grain with the aim to obtain an optimal flour yield of 70-75%, and the preparation of binary mixtures of composite types of flour with specified technological properties, there was analyzed the samples' quality in terms of molecular weight (MM). The analysis was carried out to determine the protein molecular mass of wheat grain, flax seeds and wheat flour on the basis of the developed TSEC technique using polyethylene glycols (PEGs) as standards and marker proteins to identify the analysis results.

Molecular weight is one of the most important physicochemical constants of protein. It serves to identify each protein, to understand the processes in which its functional activity is manifested. The possibilities for determining the molecular characteristics of biopolymers are significantly expanded in the presence of a linear dependence of the retention volume (Vi) on the logarithm of the biopolymer MM (lgMi). Unlike nonlinear, this relationship allows considering the chromatogram as a mirror image of the molecular weight distribution (MWD) of the analyzed sample on a logarithmic scale. As a result, the processing of chromatographic data is simplified and accelerated and it becomes possible to calibrate the chromatographic system, which improves the interpretation of the chromatogram in MWD [6,7].

Glass plates with dimensions of 100x200 mm were used for the analysis. The test samples were applied in an amount of 1-5 μl (100-500 μg). The sorbents used are Sephadexes G - 50, G - 75, G - 100, G - 150, G - 200 of the "Superfine" brand. The obtained chromatograms were taken on chromatographic paper Filtrak FN PN 7, Filtrak FN PN 3 (Germany) as a paper replica. Elution systems used: distilled water (pH 7.0), weak saline solutions (8% NaCl); phosphate-buffered solutions (pH 8.5, 8.0, 6.86, 4.5). Optimization of the process was carried out using the angle of inclination (β) varied in the range of 12-18o. The quantitative assessment was carried out on a Den-Scan densitometer (LenChrom Company) and a Shimadzy CS-920 scanner. The criterion for choosing the optimal separation parameters was the process speed (U) and the slope of the calibration curve (β) in conjunction with the sorbent grade.

The study of the grinding process was carried out on a laboratory mill unit PCA with grooved rollers for break systems (P-10 1/cm) and rollers with a micro-rough surface for grinding.

IOP Conf. Series: Materials Science and Engineering 1227 (2022) 012010 doi:10.1088/1757-899X/1227/1/012010

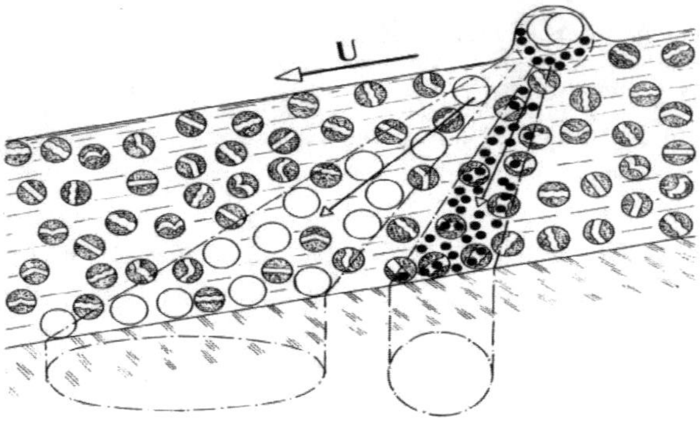

Figure 1. General view of the process of molecular weight distribution (MWD) of biopolymers and the formation of chromatographic zones in TSEC.

Sample preparation of marker proteins was carried out by dissolving in distilled water (0.2 mg of protein in 0.05 ml of water) with the addition of a few drops of 0.1 HCl until a clear solution was obtained. Proteins with known molecular weights were used: chain B insulin (MW 6000), ribonuclease (MW 13 700), chymotrypsin (MW 25 000), egg albumin (MW 43 000), and bovine serum albumin (MW 68 000).

The results of experimental studies of standard proteins are presented in Figures 2 and 3. The molecular weights of the studied proteins were determined by the ratio dependence of the analyte path length on the molecular weight logarithm (logMM).

Graphs of the dependence of the path length of proteins on the logarithm of MW are shown in Figures 4 and 7.

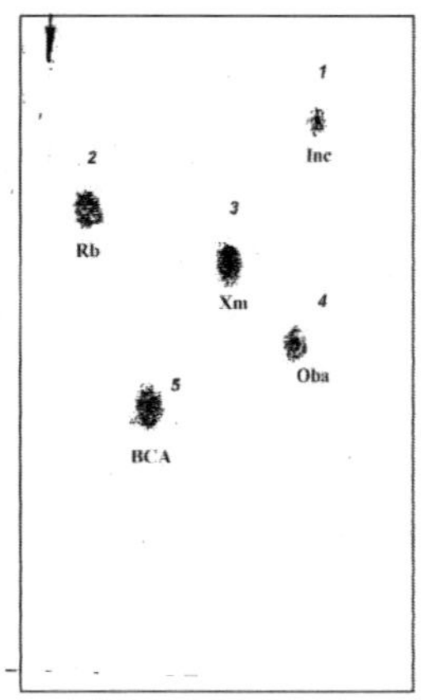

Figure 2. Chromatogram of standard proteins on G-100Sf. Parameters: pH=8.0, β=15, t=150 min, 1- insulin, 2 - ribonuclease, 3 - chymotrypsin, 4 - ovalbumin, 5 - bovine serum albumin.

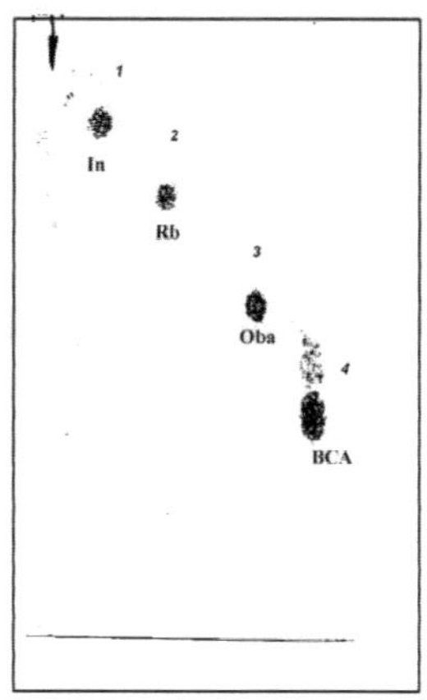

Figure 3. Chromatogram of standard proteins on G-200Sf. Parameters: pH=8,````````````````````б]] 0, β=15, t=150 min, 1 - insulin, 2 - ribonuclease, 3 - ovalbumin, 4 - bovine serum albumin.

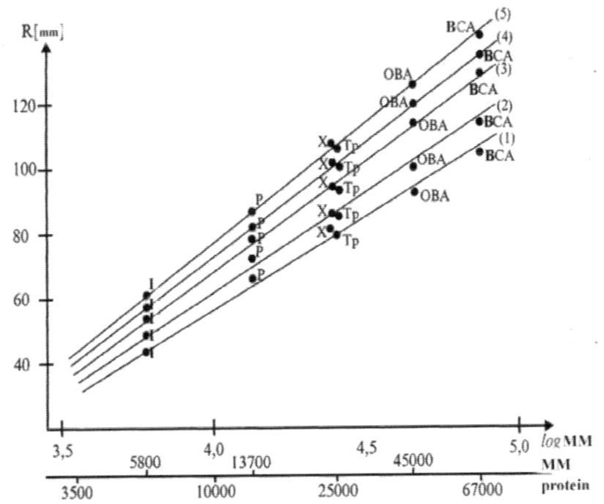

Figure 4. Dependence of the path length R on the logarithm of the MW of standard proteins on Sephadex G-75 "Superfine". Elution time 180 min. The tilt angle of the camera varied: (1)-100, (2)-120, (3)-150, (4)-180, (5)-200.

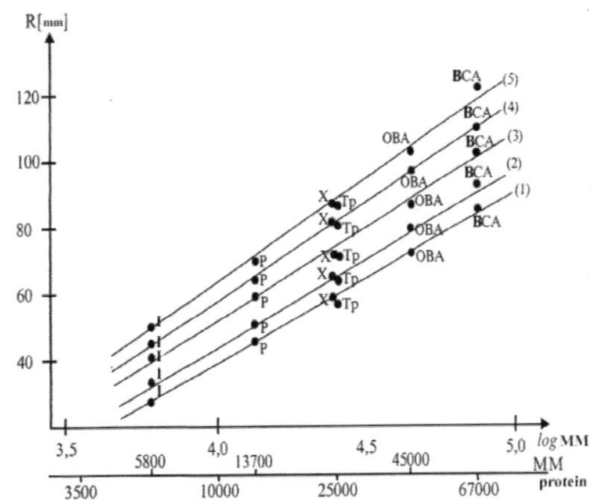

Figure 5. Dependence of the path length R on the logarithm of the MW of standard proteins on Sephadex G-100 "Superfine". Elution time 180 min. The tilt angle of the camera varied: (1)-100, (2)-120, (3)-150, (4)-180, (5)-200.

Whole proteins. It is known that dilute alkali solutions dissolve gluten, as well as extract gluten proteins directly from wheat meal and flour. The whole proteins of wheat grain, flax seeds and wheat flour were extracted with 0,2% sodium hydroxide (NaOH) solution in 50% ethanol. As a result of extraction for an hour with a 0,2% sodium hydroxide (NaOH) solution in 50% ethanol, it is possible to transfer 94-98% of all grain protein into a solution [8].

To determine the molecular mass of whole proteins of wheat grain, the obtained extract was subjected to separation on Sephadexes G–50 (Fig. 8), G–75, G–100, G–200. When analyzing the samples, eight fractions with a molecular weight of 6 300 to 84 000 were found when eluted with a phosphate buffer solution (pH 6,86). Elution time 1 hour 10 minutes Slope angle - 12°. Elution with distilled water for the same time yielded fractions with molecular weights ranging from 5 600 to 80 000.

Separation of proteins on Sephadex G-75 and elution in phosphate buffer solution (pH 4,5) yielded 6 fractions with molecular weights from 7 000 to 35 000. Elution time was 1 hour and 30 minutes. Elution with distilled water yielded 7 fractions with molecular weights ranging from 9 000 to 51 000. Elution time was 2 hours 10 minutes. Tilt angle 15°.

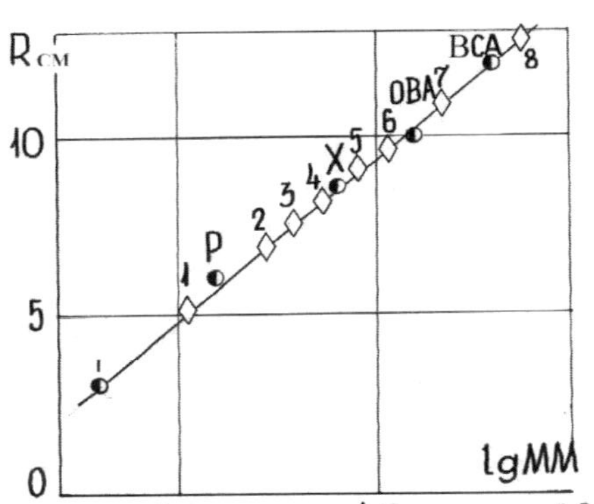

Figure 6. Dependence of the path length on the logarithm of the MW of the whole proteins of wheat grain analyzed on G-50; pH – 6.86, β=15⁰, t = 70 min.

IOP Conf. Series: Materials Science and Engineering 1227 (2022) 012010 doi:10.1088/1757-899X/1227/1/012010

The study of total proteins showed: wheat grain on Sephadex G-100 included 10 fractions with MW from 5 500 to 81 000; flour - 9 fractions with MW from 5 200 to 75 000; flax seeds - 7 fractions with MW from 6 000 to 98,000 when eluted with phosphate buffered saline (pH 8.0). Elution time - 2 hours 30 minutes, tilt angle 18°. When eluted with distilled water, fractions with MW from 5 000 to 48 000 were obtained. Elution time was 2 hours 15 minutes. Tilt angle 15 [9,10].

Separation of wheat grain proteins on Sephadex G - 200 obtained 8 fractions with a molecular weight of 10 500 to 76 000 in phosphate buffered saline (pH 6,86). The separation time was 5 hours at an angle of inclination of 18°, and under the same conditions of chromatographic separation in distilled water, 5 fractions with molecular weights from 5 500 to 57 000 were obtained.

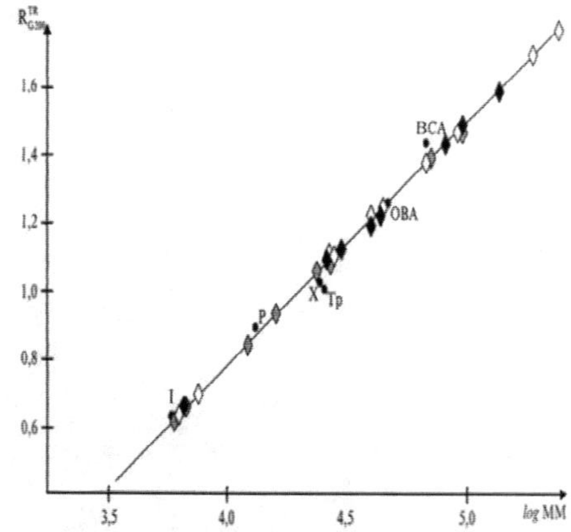

Figure 7. Dependence of the relative path length on the logarithm of the MM of whole proteins of wheat grain, flax seeds and flour on G-100, β=180, pH – 8,0.

◊ - whole proteins of wheat grain (10 fractions)

♦ - whole proteins of flour (9 fractions)

♦ - whole proteins of flax seeds (7 fractions)

Fractional proteins. There was carried out extraction of fractional proteins of wheat grain and wheat flour (2 g of a dry sample) according to the modified Osborne scheme: albumins were extracted with distilled water (6 ml), globulins - 0.5 M NaCl solution (3,5 ml), prolamins - 70% ethanol solution (4.0 ml), glutelins – 0.1 M acetic acid solution (4,0 ml) or 0.2% sodium hydroxide solution (3,5 ml). The resulting solution was purified from starch by centrifugation (5000 rpm), the resulting centrifuge served as a material for research [11,12]

Analysis of fractional proteins of wheat and flour of the premium grade showed the following results. On Sephadex G-50, there were obtained 4 albumin fractions, their molecular weights range from 8 000 to 21 500; 6 fractions of globulins, with a molecular weight from 10 000 to 32 000; 6 fractions of prolamins (gliadins), with molecular weights from 19 800 to 78 000; 4 fractions of glutelins (glutenins), with molecular weights from 40 000 to 100 000. Elution was carried out in phosphate buffered saline (pH 4,5). The separation time was 1 hour 15 minutes at an angle of inclination of 12° [13].

On Sephadex G-150, under the same conditions for protein extraction from flour, the results were: 5 fractions of albumin with a molecular weight from 6 200 to 26 000; 5 fractions of globulins with a molecular weight from 7 000 to 28 500; 6 fractions of gliadins with molecular weights from 18 500 to 56 000 and 4 fractions of glutelins with molecular weights from 40 000 to 100 000. Duration of separation is 2 hours 30 minutes. Tilt angle 15°.

Chromatographic analysis of wheat flour proteins of different solubility was carried out on Sephadex G-50 and G-100 in phosphate buffer solution (pH 6.86). Sephadex G - 50 yielded 4 albumin fractions with molecular weights ranging from 9 000 to 21 500; 5 fractions of globulins with molecular weight from 10 000 to 31 000; 7 fractions of gliadins with a molecular weight of 18 000 to 76 000; 4 fractions

IOP Conf. Series: Materials Science and Engineering 1227 (2022) 012010 doi:10.1088/1757-899X/1227/1/012010

of glutelins with a molecular weight from 40 000 to 100 000. The separation time was 1 hour 15 minutes, tilt angle 15°.

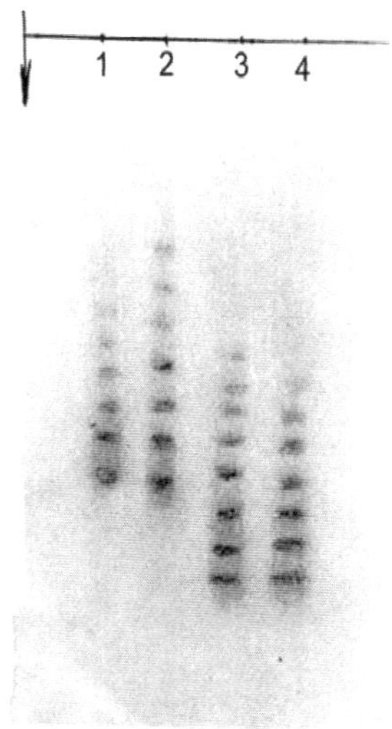

Figure 8. Chromatogram of fractional proteins of wheat flour of the premium grade for G-150. pH 6.86, β = 12°, t = 65 min. Option 3.

Discussion of a large amount of material on the analytical study of whole and fractional proteins of wheat grain, wheat flour and flax seeds is based on the main conclusion that all the studied samples have the same character in terms of fractional composition - in a qualitative sense. The differences lie in the quantitative ratio of fractions.

In whole proteins, there were found from 8 to 10 fractions, and when using differentiating solvents, up to 35 fractions were extracted. Thus, it is obvious that the TSEC method can replace a set of 2-3 other known methods of protein analysis and give a complete qualitative description of the samples under study.

The results of the quantitative analysis presented in Table 2 showed that there is a noticeable similarity in the assessment of the proportions of high molecular weight proteins in wheat and wheat flour and a noticeable difference in the proportion of these fractions in flax seeds (about 10%).

Table 1. Quantitative distribution of protein fractions of wheat, wheat flour and flax seeds.

№	Molecular weight	Wheat flour, %	Wheat, %	Flax seeds, %
1	6 000 – 15 000	5.9	6.8	6.7
2	16 000 – 30 500	7.5	9.6	11.2
3	31 000 – 40 500	9.2	11.9	15.6
4	41 000 – 58 000	15.6	16.7	18.9
5	59 000 – 78 000	25.0	22.7	20.8
6	80 000 – 100 000	36.8	32.5	25.8

The estimation of measurement errors was carried out on the basis of the standard deviation (for each series of measurements). The absolute error was determined taking into account the student's coefficient at the selected level of confidence P = 0.95.

Table 3 presents the results of statistical processing of wheat grain data. Polyethylene glycols (PEGs) were used as a test material. The molecular weights of PEGs are determined by their synthesis conditions and are reference characteristics. Statistical processing of the experimental results was carried out in relation to direct experimental data in primary units of chromatographic path lengths. The data of the qualitative analysis (by the molecular mass of grain and flour proteins) were processed according to the materials of the final estimates based on the calibration curves and are expressed in units of molecular weight.

Summarizing the results of statistical processing, it can be stated that the relative error is maximum for relatively low molecular weight components, where it can reach 20% (for soluble proteins such as albumin and globulins) and below 10% for high molecular weight fractions (proteins such as prolamins and glutelins). Quantitative analysis, the error of which is no longer due to the chromatographic procedure itself, but precisely to the method of determination, is much more accurate and gives an error in the range of 2 - 5%.

Table 2. Determination of the protein composition of wheat grain (G - 50).

№	Average molecular weight	Standard deviation S	Relative error Sr, %	Confidence interval +-Δ
1	8 500	500	5.8	620
2	13 500	400	2.9	490
3	16 900	400	2.3	490
4	24 000	750	3.1	930
5	32 000	750	2.3	930
6	45 600	690	1.5	850
7	56 200	520	1.0	640
8	64 900	840	1.3	1040
9	81 480	800	1.0	900
10	98 700	1300	1.3	1600

4. Conclusion

The study allowed us to reveal the regularities of the preparation and grinding of binary grain mixtures and then there were modeled the main mixing methods to obtain composite types of flour with given technological properties and increased nutritional value. Based on the analysis of the results of grinding the grain mixture by coarse systems of the break process, a technological scheme of grinding was developed, which ensured the flour yield of 70 - 75%.

Detailed analytical studies of plant proteins for obtaining qualitative and quantitative characteristics by the TSEC method are presented in the paper. Analytical characteristics of the method have been studied, which makes it possible to obtain reproducible results of a kind of qualitative analysis of biopolymers by quantitative characteristics of molecular weights and quantitative analysis of proteins by optical densitometry. The developed methods for determining the molecular mass of proteins are characterized from the metrological point of view.

References

[1] Makarov Yu I 2019 Apparatus for mixing bulk materials *Mechanical Engineering* 190-4
[2] Chang R 2010 Physical chemistry with applications to biological systems *World 116-8*
[3] Witkiewicz L, Mazurek H, Bladek J 2018 Chromatography Planar *World* 121-2.
[4] Tomasek V 2020 Laboratory manual for chromatographic and related methods *Institute of Organic Chemistry and Biochemistry of the Academy of Sciences* 22-6
[5] Kozmina N P 2016 Biochemistry of grain and products of its processing *Kolos* 21-2
[6] Kretovich V L 2018 Biochemistry of plants Higher school 65-6.
[7] Gurkovskaya E A 2012 Investigation of cereals' macromolecular composition by chromatographic methods *KhlebProdInform* 15-20
[8] Gurkovskaya E A 2003 Investigation of protein complexes of cereals by thin-layer exclusion chromatography. Low molecular weight proteins. High molecular weight proteins *KhlebProdInform* 3-29.
[9] Kechkin I, Ermolaev V, Romanenko A, Ivanov M, Gurkovskaya E 2020 Dependence of fat acidity value on wheat grain storage conditions *International Conference on Food Industry* 34-9
[10] Kechkin I, Ermolaev V, Romanenko A, Tarakanova V, Buzetti K 2020 Management of air flows

inside steel silo during grain storage *International Conference on Food Industry, Economy and Security* 22-5
[11] Huang Y, Li C 2017 Real-time monitoring system for paddy environmental information based on DC powerline communication technology *Computers and Electron in Agricult* 51–62
[12] Stone G, Glover D 2017 Disembedding grain: Golden Rice, the Green Revolution, and heirloom seeds in the Philippines *Agricult and Human* 87–102
[13] Bhatta M, Belamkar V, Baenziger P, Morgounov A Genome-wide association study reveals novel genomic regions for grain yield and yield-related traits in drought-stressed synthetic hexaploid wheat *Int. J. of Molecular Sci* **19(10)** 92-9

MIST Aerospace-IV 2021 IOP Publishing
IOP Conf. Series: Materials Science and Engineering 1227 (2022) 012011 doi:10.1088/1757-899X/1227/1/012011

Investigating the dynamics of a buck converter with time-delay feedback control

A I Andriyanov

Department of Electronic, Radioelectronic and Electrotechnical Systems, Bryansk State Technical University, Bulvar 50 Let Oktyabrya, 7, Bryansk, 241035, Russia

E-mail: mail@ahaos.ru

Abstract. The paper deals with the dynamics features of a closed automatic control system based on a buck converter with a function to control nonlinear dynamic processes. A comprehensive analysis of the system under study is carried out using both small-signal average dynamic models and nonlinear dynamic models. For the first time, a small-signal average model of a system with a control function for nonlinear dynamic processes based on a time-delay feedback control has been obtained. The influence of this control method on Bode plots of an open system is shown. It is noted that the use of the time-delay feedback control ensures the stability of both the equilibrium point of the small-signal average model and the fixed point of the periodic mode of the nonlinear dynamic model of the system under consideration. This enables to achieve high quality output voltage. The application of the time-delay feedback control does not require to correct the system parameters, which ensures a given speed and low losses in the power switch of the converter. In addition, the quality of the transition process in the system is improved, namely, oscillation is eliminated and overshoot is reduced. Using the time-delay feedback control, it is possible to provide the desired dynamic mode in the areas of instability of the initial system without controlling nonlinear dynamic processes, which makes it possible to increase the speed due to the small time constant of the regulator. These results were obtained for the first time and can be used in the design of pulse power supply systems of a wide variety.

1. Introduction

Pulsed-width converters of electric power [1, 2] are actively used in various fields of activity. They are closed-loop automatic control systems that require stability at the design stage. This is done by choosing the optimal parameters of the regulator based on the required stability margin in phase and/or amplitude when considering the system as linear.

Small-signal average models of pulse voltage converters are known to describe electromagnetic processes only approximately, since these converters belong to the class of nonlinear dynamic systems. Dynamic nonlinearities lead to the fact that undesirable oscillations without loss of stability of the equilibrium point of a small-signal average model of the system can appear at the output of the converter as a result of bifurcations with variations of one or more of its parameters [3]. A complex analysis of such phenomena is possible only with the use of nonlinear dynamic models of systems of the class under consideration, the theory of bifurcations and chaos, as well as the theory of Poincare maps [4-6].

When analyzing pulse converters as nonlinear dynamic systems, the concept of the desired dynamic mode is used. The desired dynamic mode is understood as a mode in which the frequency of oscillations of the output voltage is equal to the frequency of pulse-width modulation (*PWM*) (1-cycle) [4]. Also, as a result of bifurcations, *m*-cycles can be realized, where *m* is the cycle ratio, which shows how many times the frequency of fluctuations in the output voltage is less than the frequency of pulse-width modulation. In addition, chaotic oscillations ($m = \infty$) may be observed during the operation of pulse converters. The operation of pulse converters in undesirable modes is usually accompanied by a large amplitude of output voltage fluctuations [3-6].

Elimination of undesirable dynamic modes is currently possible both with the help of parametric synthesis and with the help of structural and algorithmic synthesis. Parametric synthesis is associated with the correction of the controller parameters calculated when considering the system as linear, which can lead to a decrease in its performance. Increasing the switching frequency is an effective method of controlling nonlinear phenomena [7], but it is accompanied by an increase in dynamic losses in power switches, which makes it impossible to use it when using low speed switches. Also, increasing the frequency of power switches may not be possible due to the limited speed of the microcontroller.

The most promising approach to eliminating undesirable dynamic modes is structural and algorithmic synthesis. In this case, it is assumed to make a specific structure of the control system, which eliminates unwanted fluctuations in the output voltage. At the same time, the correction of the parameters of the controller or the power section is not carried out, which gives the desired speed, as well as acceptable dynamic losses.

The most promising and simple method of controlling nonlinear dynamic processes of pulse power supply systems is the time-delay feedback control (*TDFC*) [8-10]. Here, an additional control system for nonlinear dynamic processes (*NDPCS*) is introduced into the control system with a standard structure, which supplies a corrective effect to the control circuit of the main control system at the beginning of each clock interval [11]. This allows to change the properties of a nonlinear dynamic system and ensure the stability of the desired dynamic mode. The periodic mode is stable when its largest multiplier lies within the unit circle on the complex plane. The analysis of the local stability of periodic modes in this case is carried out on the basis of the first Lyapunov method [4].

Existing works devoted to the application of the time-delay feedback control, as a rule, show the possibility of stabilizing the desired dynamic mode, but at the same time a number of issues are overlooked. For example, how does the application of the time-delay feedback control affect the transient response of the system or how does Bode plot of an open small-signal average model of the system change? Also, in paper [12] the Bode diagrams of the continuous control system based on *TDFC* were considered. At the same time, when considering microprocessor control systems, it is necessary to consider a discrete control system.

In this paper, a small-signal average dynamic model of a buck converter with a discrete control system based on the time-delay feedback control is proposed for the first time, and Bode plots of an open-loop system are built. The influence of the time-delay feedback control on the dynamics of the system from various points of view is also studied.

2. Description of a closed automatic control system based on a buck converter with a control function for nonlinear dynamic processes

The functional diagram of the automatic control system (*ACS*) based on a buck converter with a proportional-integral regulator (PI-regulator) is shown in figure 1, a. The following designations are used in the figure: *GCS* is the general control system, *NDPCS* is the nonlinear dynamic processes control system, *VT* is a power transistor, *VD* is a power diode, R_{LD} is the load resistance, *L* is the filter choke, R_L is the active resistance of the filter choke, *C* is the filter capacitor, R_c is the active resistance of the capacitor, β is the feedback amplifier with β coefficient, K_{r1} is the gain coefficient of the proportional part of PI-regulator, *SH* is a sample-and-hold device, *MC* is a master clock, *RG* is a ramp generator, «= =» is a comparator, SH_i are sample-and-hold devices of the scaled value of *i*-phase variable with β_i coefficient, CB_i is the control block for *i*-phase variable, K_i is the gain coefficient of CB_i output signal, U_{in} is an input voltage, U_{mc} is the output signal of the master clock, U_{rmp} is the output signal of the ramp generator, U_{ref} is the reference signal, U_{err} is the error signal, U_{con} is the control signal, U_p are power switch control pulses, $W_{i(s)}$ is the integrator transfer function

IOP Conf. Series: Materials Science and Engineering 1227 (2022) 012011 doi:10.1088/1757-899X/1227/1/012011

$$W_i(s) = \frac{1}{T_i s + K_{r2}},$$

where T_i is the integrator time constant, K_{r2} is the coefficient of stray parameters of the integrator circuit [13].

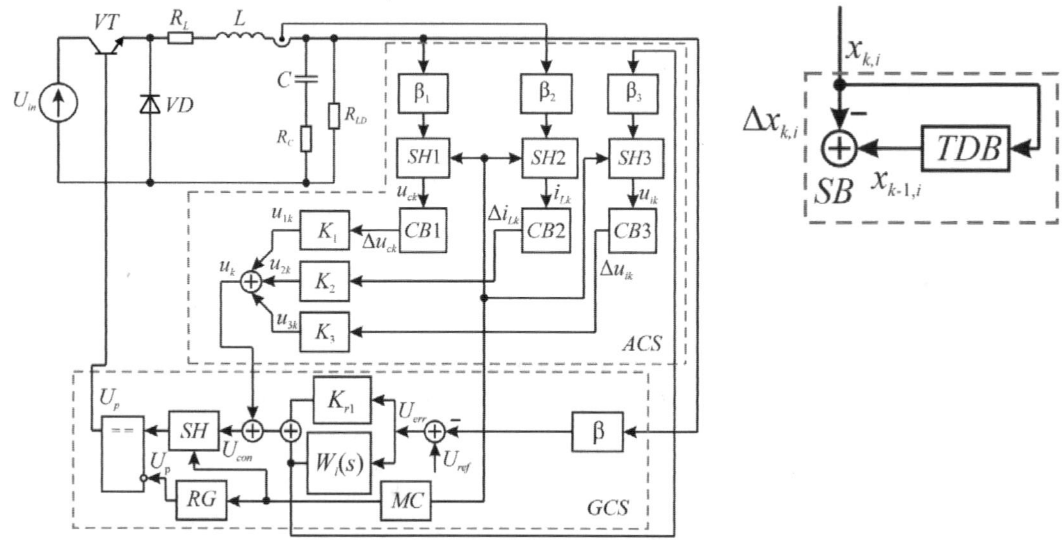

(a) (b)

Figure 1. Description of *ACS* operation: a - *ACS* block diagram, b - *CB* block diagram.

The green frame in figure 1, a outlines the general control system which aims at stabilizing the average value of the output voltage of the converter. The red frame outlines *NDPCS*, which is discrete and gives corrective actions with the frequency equal to the pulse-width modulation frequency, which allows to stabilize the desired 1-cycle.

Figure 1, b presents a block diagram of CB_i correcting block. As the figure shows, this block finds the difference between *i*-phase variable at the beginning of the current *k*-clock interval ($x_{k,i}$) and the previous *k*-1 clock interval ($x_{k-1,i}$)

$$\Delta x_{k,i} = x_{k-1,i} - x_{k,i}, \text{ where } i=1, 2, 3.$$

Then the residual for each phase variable is scaled with K_i coefficient. The corrective effect of *NDPCS* for the system in question is defined by the expression

$$u_k = \sum_{i=1}^{3} K_i \Delta x_{k,i}.$$

It is known that the main task of controlling a nonlinear dynamic process is to ensure $u_k=0$ in steady-state mode when the system operates in the desired 1-cycle $x_{k-1,i}=x_{k,i}$.

3. Small-signal linearized dynamic model of the system

The structure of the small-signal dynamic model of the system (figure 1) is shown in figure 2, where the following designations are used: $W_{pm}(s)$ is the transfer function of the pulse modulator, $W_{du}(s)$ is the transfer function of the duty cycle-output, $W_{di}(s)$ is the transfer function of the duty cycle–throttle current, $W_{cb}(s)$ is the transfer function of the CB_i correcting blocks, $u_{out} \approx u_c$ is the output voltage, u_c is the voltage on the capacitor, i_L is the throttle current, u_i is the output voltage of the integrator as part of PI-regulator. The rest of the designations correspond to figure 1.

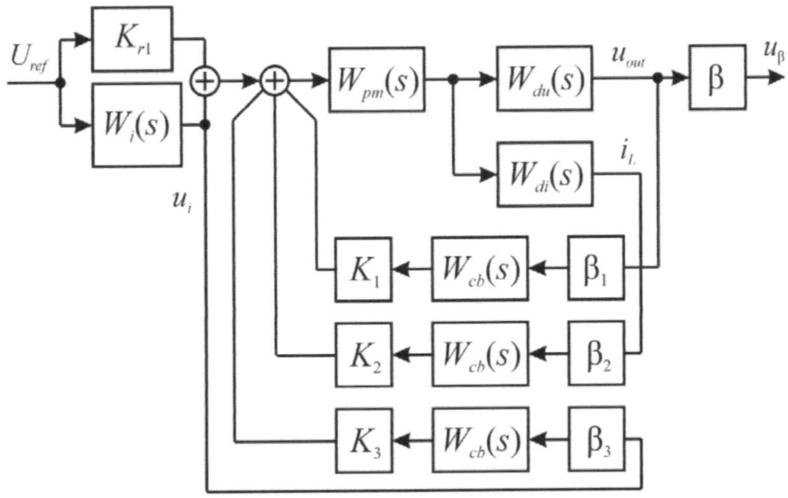

Figure 2. Small-signal average model of the system with controlling nonlinear dynamic processes.

The transfer function of PI-regulator, taking into account its imperfection, is determined by the expression [13]

$$W_r(s) = K_{r1} + W_i(s) = K_{r1} + \frac{1}{T_i s + K_{r2}}.$$

The transfer function of the duty cycle-output is defined as [1]

$$W_{du}(s) = \frac{\hat{u}_{out}(s)}{\hat{d}(s)} = \frac{DR_{LD} + CDR_{LD}R_C s}{LC(R_{LD}+R_C)s^2 + (L + R_{LD}R_L C + R_{LD}R_C C + R_C R_L C)s + R_{LD} + R_L},$$

where D is the duty cycle, $\hat{d}$ is the small perturbation of the duty cycle, $\hat{u}_{out}$ is the small perturbation of the output voltage.

The transfer function of the duty cycle–choke current is defined as

$$W_{di}(s) = \frac{\hat{i}_L(s)}{\hat{d}} = \frac{U_{in} + CU_{in}s(R_L + R_C)}{LC(R_{LD}+R_C)s^2 + (L + R_{LD}R_L C + R_{LD}R_C C + R_C R_L C)s + R_{LD} + R_L}.$$

The transfer function of the pulse modulator is as follows

$$W_{di}(s) = \frac{1}{U_{rmp,m}},$$

where $U_{rmp,m}$ is the amplitude of the sweeping voltage.

Let us get the transfer function $W_{cb}(s)$ (figure 1, b). As it is seen from the figure, *CB* block is described by a difference equation

$$\Delta x_{k,i} = x_{k-1,i} - x_{k,i}.$$

Using z-transformation, we have

$$\Delta x_{k,i} = x_{k,i}\left(z^{-1} - 1\right).$$

The discrete transfer function of *CB* block is the following

$$W_{cb}(s) = -1 + z^{-1}, \tag{1}$$

Using a Tustin's method, we proceed to a continuous transfer function. To do this, we put the following expression in (1) [14, 15]

$$z = \frac{1 + \frac{T}{2}s}{1 - \frac{T}{2}s},$$

where T is the duration of the clock interval *PWM*.

Thus, the continuous transfer function of CB_i blocks looks as follows

$$W_{cb}(s) = \frac{2Ts}{Ts + 2}.$$

4. Simulation of system dynamics

This section gives the results of modeling a closed automatic control system based a buck converter. The simulation was carried out with the following system parameters: input voltage U_{in}=800 V, choke inductance L=0.009 H, active resistance of the choke R_L=10 Ohm, capacitor capacity C=10 uF, load resistance R_{LD}=250 Ohm, coefficient taking into account the imperfection of PI-regulator K_{r2}=0.001, feedback factor β=0.01, amplitude of the sawtooth voltage $U_{rmp,m}$=10 V, *PWM* frequency f_{sw}=20 kHz, U_{ref}=5 V.

The experiment is carried out with a system of PI-regulator without *NDPCS* as well as a system adjusted using *NPDCS*. When modeling, we used: a nonlinear dynamic model [4] and a small-signal, average model.

Figure 3 presents the results obtained using a nonlinear dynamic model of a system without *NPDCS*, in the form of a map of dynamic modes and a corresponding diagram of the range of output voltage fluctuations. These diagrams are built in the range of two parameters of PI-regulator: K_{r1} and K_i, where $K_i=1/T_i$. On the map of dynamic modes, $D_{i,j}$ symbols mark the areas of different modes, where i symbol is the multiplicity of m-cycle, and j symbol is the number of the periodic mode with i-multiplicity on the map. So, for example, $D_{1,1}$ is the first area of the desired 1-cycle. $D_{ch,j}$ areas correspond to the chaotic modes of the converter operation ($m \rightarrow \infty$).

AB boundary is a bifurcation boundary, which crossing leads to undesirable dynamic modes (in this case, mainly chaotic). As it can be seen from figure 3, b, undesirable modes are accompanied by a large amplitude of output voltage fluctuations. *CD* boundary highlighted in red represents the stability boundary of a small-signal average model of a system without *NPDCS* with a phase stability margin $\varphi=10^{\circ}$. This boundary was calculated according to the method [16]. It is obvious that the stability boundary of the desired 1-cycle (*AB*) and the stability boundary of the small-signal average *CD* model do not coincide, which indicates the need to use not only small-signal average models at the design stage, but also nonlinear dynamic models that allow taking into account the possibility of nonlinear oscillations.

 IOP Publishing
doi:10.1088/1757-899X/1227/1/012011

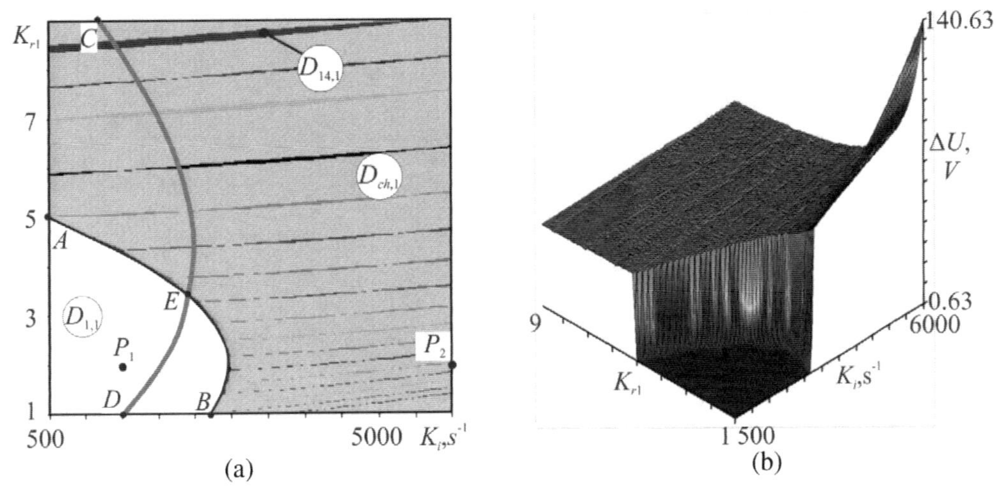

Figure 3. Two-parameter diagrams: a - the map of dynamic modes, b - the diagram of the range of output voltage fluctuations.

Figure 3 shows that the desired area of the regulator parameters is the area bounded by *AED* curve, which is determined by the intersection of the stability area of the 1-cycle and the stability area of the linearized system model. In the area bounded by *DEB* curve, as follows from figure 3, a, the desired 1-cycle is observed, but the small-signal average model of the system in this case is stable with a stability margin of $\varphi < 10^{\circ}$, so we consider the operation in this area unacceptable.

Using the time-delay feedback control, it is possible to ensure the operation of the system in the desired dynamic mode over the entire range of the regulator parameters (figure 3, a) [7]. It is obvious that *NDPCS* use changes Bode plot of a small-signal average system model and, accordingly, affects transients. Let us evaluate these indicators with the following *NDPCS* parameters: K_1=9.62334, K_2=4.74367, K_3=0.190714, β_1=-0.0214938, β_2=0.0117646, β_3=-0.836231.

The calculation results are shown in figure 4 as Bode plots and their corresponding time diagrams. Here, diagrams built for the system without *NDPCS* are highlighted in red, and for the system using *NDPCS* in blue. Bode plots are built using a small-signal average model, and time diagrams using a nonlinear dynamic model. As we can see from the figure, the use of *NDPCS* results in the transformation of diagrams. So, in particular, at P_1 point, where the small-signal average model of the system without *NDPCS* is stable, the use of *NDPCS* allowed to increase the stability margin from 22° to 67° (figure 4, a), overshoot was also excluded (the time diagram of the system with *NDPCS* is highlighted in blue in figure 4, c).

At P_2 point, the small-signal average model of the system without *NDPCS* is unstable, which is accompanied by fluctuations in the output voltage (highlighted in red in figure 4, d). The phase at the edge frequency of the open system in this case is 184°, i.e. more than 180°. The use of *NDPCS* at P_2 point made it possible to provide a stability margin of 33°, which eliminated fluctuations in the output voltage (highlighted in blue in figure 4, d).

The analysis of diagrams in figure 4 shows that the use of *NDPCS* does not lead to their significant transformation, which provides a speed capacity close to what was defined when choosing the parameters of PI-regulator (figure 4, c). This ensures the stability of both the equilibrium point of the linear system and the fixed point of the desired dynamic mode of the nonlinear dynamic system. It is also obvious that when selecting the controller parameters corresponding to P_2 point, the system speed capacity increases significantly (figure 4, d), while taking into account the use of *NDPCS*, it remains stable.

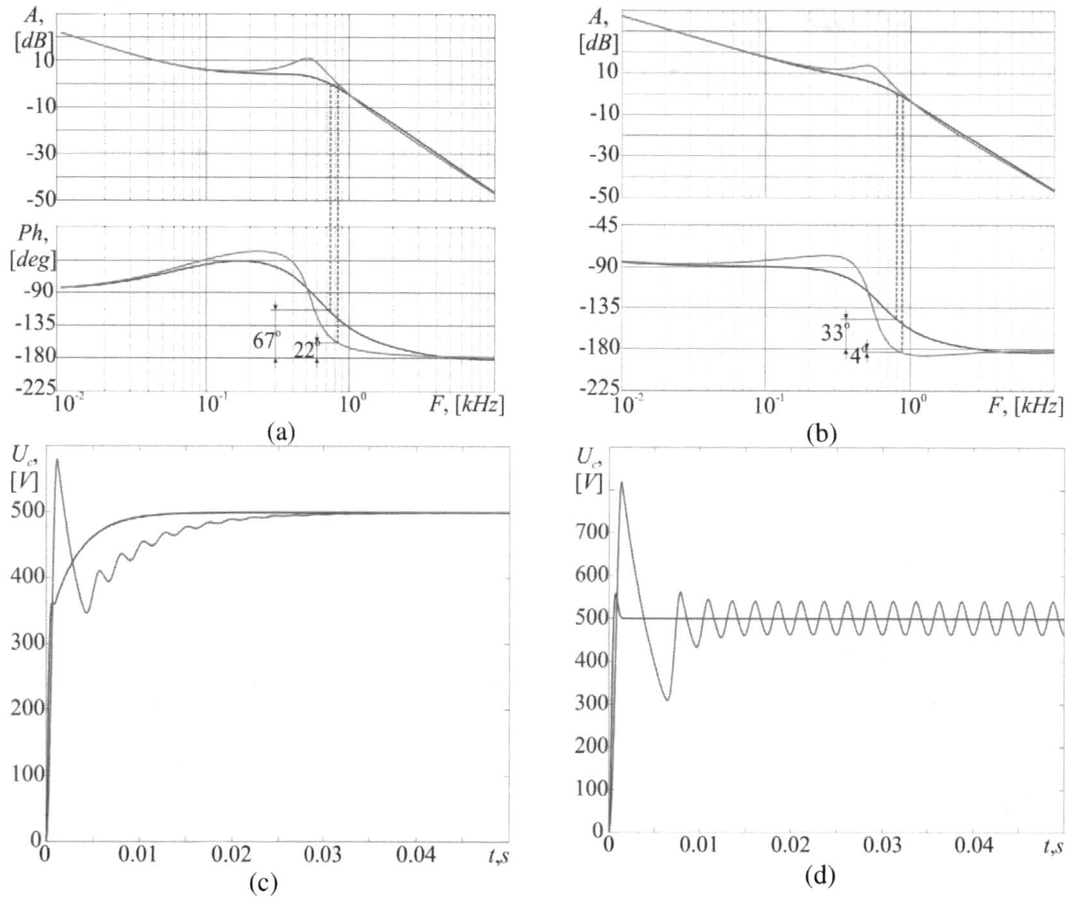

Figure 4. Diagrams for *ACS* with buck converter: a - Bode plots at P_1 point, b - time diagrams at P_1 point, c - Bode plots at P_2 point, d - time diagrams at P_2 point.

5. Conclusion

According to the research results, the following conclusions can be drawn.

1. For the first time a small-signal linearized model of a system with a control function for nonlinear dynamic processes based on *TDFC* was obtained, which allows analyzing frequency characteristics.
2. It is shown that the use of *TDFC* even at the point of stability of an uncorrected system makes it possible to improve the quality of the transition process, in particular, overshoot was completely eliminated while maintaining the duration of the transition process.
3. At the point of instability of an uncorrected system, the use of *TFDC* ensures its stability, which provides an acceptable quality of the transition characteristic.
4. The presented results were obtained for the first time and they allow deeper understanding of the effect of the time-delay feedback control on the dynamics of the system as a whole.

References

[1] Severns R P and Bloom G 1985 *Modern DC-to-DC Switchmode Power Converter Circuits* (New York: Van Nostrand Reinhold Company) p 361
[2] Rashid M 2010 *Power Electronics Handbook* (Butterworth-Heinemann) p 895
[3] Andriyanov A I 2021 Design of Pulse-width DC/DC Converters Taking into Account Dynamic Nonlinearity *Electrotechnical Systems and Complexes* **2** 39-44

[4] Zhusubaliyev Zh T and Mosekilde E *Bifurcations and chaos in piecewise-smooth dynamical systems* (Singapore: World Scientific Publishing Co. Pte. Ltd) p 376

[5] Zhusubaliyev Z T, Soukhoterin E A and Mosekilde E 2003 Quasi-periodicity and border-collision bifurcations in a DC-DC converter with pulse-width *modulation IEEE Trans. on Circuits and Systems I: Fundamental Theory and Applications* **50** 1047-57

[6] Zhusubaliyev Z T, Soukhoterin E A and Mosekilde E 2001 Border-collision bifurcations and chaotic oscillations in a piecewise-smooth dynamical system *Int. J. Bifurcation and Chaos* **11** 2977-3001

[7] Andriyanov A I 2021 Dynamics of a Buck Converter with PI Controller 2021 *Int. Conf. on Industrial Engineering, Applications and Manufacturing* (Sochi: IEEE) pp 366-71

[8] Kapitaniak T 1996 *Controlling Chaos - Theoretical and Practical Methods in NonLinear Dynamics* (London: Academic Press) p 164

[9] Robert B, Feki M and Iu H H C 2006 Control of a PWM inverter using proportional plus extended time-delayed feedback *Int. J. Bifurcation Chaos* **16** 113-28

[10] Wei-Guo L, Ping-Ye X, Luo-Wei Z and Quan-Ming L 2010 Bifurcation control of current-mode buck converter via *TDFC Chinese Phys. Lett.* **27** 030501

[11] Andriyanov A I 2017 Comparative analysis of efficiency of time-delayed feedback for nonlinear dynamics control of composite DC-to-DC power converters 2017 *Int. Conf. on Industrial Engineering, Applications and Manufacturing* (St. Petersburg: IEEE) pp 1-7

[12] Rodriguez E, Alarcón E, Iu H H C and El Aroudi A 2015 A Frequency Domain Approach for Controlling Fast-Scale Instabilities in Switching Power Converters *Int. J. Bifurcation Chaos* **25** 1550141

[13] Andriyanov A I and Sachenko E A 2016 Mathematical model of a three-phase transistor controlled rectifier with a PI voltage controller in regeneration mode 2016 *2nd International Conference on Industrial Engineering* (Chelyabinsk: IEEE) pp 1-5

[14] Unsal D and Dogan M 2014 Implementation of identification system for IMUs based on Kalman Filtering 2014 *IEEE/ION Position, Location and Navigation Symp* (Monterey, CA, USA: IEEE) pp 236-40

[15] Makarov M, Grossard M, Rodriguez-Ayerbe P and Dumur D 2016 Modeling and Preview H_∞ Control Design for Motion Control of Elastic-Joint Robots With Uncertainties 2016 *IEEE Trans. Ind. Electron* **63** 6429-38

[16] Garg M M, Hote Y V, Pathak M K and Behera L 2018 An Approach for Buck Converter PI Controller Design Using Stability Boundary Locus 2018 IEEE/PES *Transmission and Distribution Conference and Exposition* (Denver: IEEE) pp 1-5

Improving the environmental safety of vehicle operation by using flywheel batteries

L R Abdullina[1], K R Rusanova[2], M M Petrova[2], M K Avakian[3] and A V Zrazhevskiy[4]

[1]Department of Theory of machines and mechanisms, Bauman Moscow State Technical University (BMSTU), ul. Baumanskaya 2-ya, 5/1, Moscow, 105005, Russian Federation
[2]Department of Ecology and Industrial Safety, Bauman Moscow State Technical University BMSTU, ul. Baumanskaya 2-ya, 5/1, Moscow, 105005, Russian Federation
[3]Department of The Refrigerating, Cryogenic Technics, Central Airs and Life-support, Bauman Moscow State Technical University BMSTU, ul. Baumanskaya 2-ya, 5/1, Moscow, 105005, Russian Federation
[4]Platov South-Russian State Polytechnic University (NPI), Technology of mechanical engineering, technological machines and equipment, st. Prosveshcheniya, 132, Novocherkassk, Rostov region, 346428, Russian Federation

E-mail: abdullina_98@mail.ru

Abstract. Despite the fact that according to the largest international organization “Our World in Data”, air pollution mortality is decreasing - mainly due to improved indoor air - this type of environmental pollution is one of the world's leading risk factors. In 2017 approximately 5 million people around the world died from diseases caused by air pollution – strokes, lung cancer, etc. The prevailing part (about 40% of the world's air pollution) occurs when vehicles are used, mainly personal vehicles, which means that it is necessary to develop measures to improve the environmental friendliness of the use of vehicles. Since the exhaust gases of automobile engines contain a large amount of air pollutants, it is possible to limit their emissions by developing and implementing hybrid vehicles with flywheel energy accumulators. The article considers the dimensionless dependence of the current acceleration of the car to its maximum acceleration in the presence and absence of a flywheel in the system, and checks the reliability of the dependence on the stand.

1. Introduction

At the moment, in accordance with new environmental standards and changes in legislation, consumers' interest in cars powered by internal combustion engines is gradually decreasing. There is an increasing number of cars that run either entirely on electricity or in a hybrid way. Hybrid cars use two sources of energy - for example, combustible petroleum products and electricity. The effectiveness of the use of hybrid vehicles has been repeatedly confirmed by multiple studies by scientists from different countries [1 - 6].

2. Main part

By considering the principle of operation of hybrid vehicles, it is necessary to point out that in such vehicles, the power units work together, transferring power to the wheels through a complex transmission. Usually, the electric motor helps the combustion engine when the car is lifting, starting its work, in the process of overtaking. The main engine speed is reduced, wear is slowed down, fuel consumption is reduced, and dynamics remains the same or improves. The principle of operation of some hybrids allows the car to start on electric traction. The best models can travel up to 50–100 km without starting an internal combustion engine - in the city they drive silently, with zero exhaust emissions [7].

There are also hybrid cars in which one axle is driven by a gasoline engine and the other by an electric one. This allows you to increase cross-country ability and improve controllability without complicating the car's layout and almost without increasing the cost of its maintenance [8].

But the active introduction of hybrid vehicles with an electric battery leads to the need to dispose of spent batteries, which has a negative impact on the environmental situation [9-12]. In their production is included a lithium bromide, which is hazardous to human health, as well as cobalt and nickel. Scientists from the American Chemical Society (ACS) have already provided evidence that the compounds contained in lithium-ion batteries are harmful to soil bacteria [13].

Recently, hybrid power plants with braking energy recuperation have become widespread in the world, which reduce energy losses by up to 50%, depending on the frequency of the braking process [14], [15]. In practice, various types of energy accumulators are used, which are most often installed between the motor and the generator, such as electrical, flywheel, mechanical and others [16], [17].

2.1. Perspective energy storage devices

Most often, energy storage devices are subdivided into electrochemical ones, which convert electrical energy into chemical energy of substances, and physical energy storage devices, which transform into mechanical energy.

Electrochemical energy storage includes capacitive storage, molecular energy storage, inductive storage, storage batteries, superconducting inductive storage, etc [18].

Capacitive energy storages are one of the most powerful types of energy sources. They are highly reliable and transfer efficiently stored energy. Low-power chargers can be used to charge them. But due to the low specific energy intensity, it is difficult to create capacitive energy storage devices with stored energy exceeding 10 MJ.

The most common batteries are lead acid. Batteries are considered to be relatively cheap. The main disadvantages are considered to be heavy weight and size, sensitivity to negative temperatures and a limited number of discharge cycles.

Nickel-cadmium batteries are more energy efficient and can also operate at low temperatures. But recently their use has been limited due to the non-environmental friendliness of production and disposal.

Lithium-ion batteries have some advantages - they have a high energy density, a large number of operating cycles. However, they have problems with the storage and operation of the battery at low temperatures and also lose their properties when charged after partial discharge of the battery.

Heat accumulators are capable of storing thermal energy and do not require frequent addition of fuel. The main disadvantages are considered to be long heating, significant dimensions and weight - this creates difficulties in their transportation and assembly.

Unfortunately, at low temperatures, typical of most regions of our country, the properties and resource of all electrochemical storage devices are significantly deteriorated, which does not allow them to be effectively used. Therefore, an important task is to develop such a drive that meets the requirements for durability, reliability and overall dimensions.

The physical storage of electricity mainly includes two types of complexes: kinetic energy storage (flywheels) and gravitational energy storage [19].

In a flywheel accumulator, energy is accumulated and stored in the form of the kinetic energy of the flywheel or its perspective design - a super flywheel, and is released in the form of mechanical rotational energy. Thus, the type of energy is conserved without conversion, which significantly increases the overall efficiency of the battery. Compared to other energy storage methods, flywheel energy storage systems have a long service life, typically over 20 to 25 years.

Often, a flywheel energy storage unit is combined with devices for converting the type of energy - hydraulic, pneumatic, electric machines, forming an energy storage system. The most widespread are energy storage systems with electric reversible machines (engine-generator) [20].

A flywheel energy storage device based on a super flywheel has one of the highest specific power indicators among existing energy storage devices. And when using modern high-strength materials, for example, graphene tapes (paper), the flywheel battery has the highest specific energy index of all storage devices.

In terms of the specific energy reserve per unit weight, the flywheel accumulator effectively competes with the electric one, differing from it in a higher service life [21]. However, electrical energy represents the greatest convenience for the consumer. Therefore, it is advisable to combine a flywheel battery with a reversible electrical machine controlled by a computer. The flywheel battery is most often installed between the motor and the generator [22].

The flywheel energy accumulator has been found to be the most efficient for use in various fields, including the transport and material handling industry (EERA (European Energy Research Alliance)). Therefore, the best option for this project is the use of a device that would provide braking energy recuperation using a flywheel energy accumulator.

During the Soviet era, according to the results of studies of city buses with hybrid units and variators under the leadership of N.V. Gulia, it was found that fuel consumption is reduced by using flywheel batteries by about half, and the toxicity of exhaust is several times. Advantages and disadvantages of using flywheel energy accumulators have been the subject of research by both Russian and foreign scientists, while increasing the efficiency of installations when using flywheels has been proven [23], [24].

2.2. A hybrid car with a flywheel energy accumulator

This article discusses the scheme of a hybrid car with a flywheel energy accumulator. The ERA international community conducted a study of energy storage methods and confirmed that it can be safely disposed of, which means that it is the most environmentally friendly among other batteries [25].

The recovery of braking energy and its storage in the flywheel is carried out by changing the gear ratio of the variator that connects the flywheel to the transmission of the machine. A dynamic model of a machine with a variable gear ratio is shown in the figure 1.

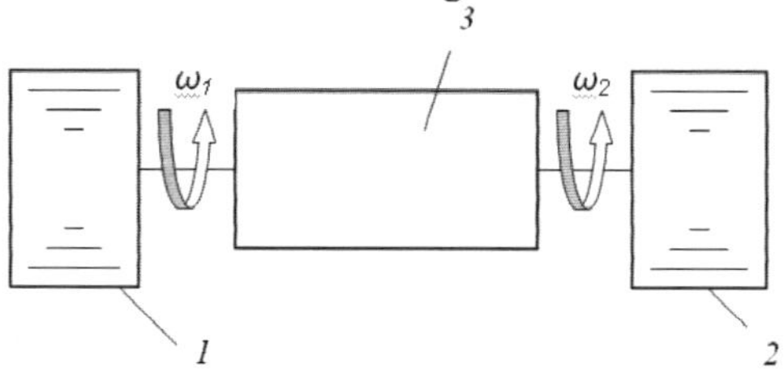

Figure 1. Block diagram of a machine with a flywheel energy accumulator: 1 - the moment of inertia of the flywheel; 2 - the moment of inertia of the transmission of the machine; 3 – the variator.

There is a need to evaluate the dynamic qualities of a machine with a flywheel energy accumulator, which functions accord to a rational dependence. This estimate can be made using the law of conservation of kinetic energy in differential form which can be written as:

$$M_{\Sigma}^{g} = J_{\Sigma}^{g}\left(\frac{d\omega_2}{dt}\right) + \left(\frac{dJ_{\Sigma}^{g}}{d\varphi_2}\right)\frac{(\omega_2)^2}{2} \tag{1}$$

where $J_{\Sigma}^{g} = J_2 + (U_{12})^2 J_1$ - the total moment of inertia of the machine, given to the transmission shaft, $U_{12} = \frac{\omega_1}{\omega_1} = U(t)$ - instantaneous gear ratio of the variator between the transmission shafts and the flywheel, M_{Σ}^{g} - the given moment of driving forces and resistance.

With the values given discounted total moment of inertia to the output shaft of the transmission of the machine after replacing variables the given value of the total moment of forces according to the law of conservation of kinetic energy in the differential form takes the form:

$$M_{\Sigma}^{g} = \varepsilon_2 \left\{ J_2 + U_{12}^2(\omega_2) J_1 + U_{12}(\omega_2) \left[\frac{dU_{12}(\omega_2)}{d\omega_2} \right] J_1 \omega_2 \right\} \quad (2)$$

Since the resulting differential equation is nonlinear, its solution was sought by numerical integration on a computer. When you perform a numerical calculation of the dynamic properties of a car with braking energy recovery, the ratio of the current acceleration ε_2 of the transmission drive shaft to its maximum value $(\varepsilon_2)_{max}$ at the beginning of acceleration:

$$\frac{\varepsilon_2(\omega_2)}{(\varepsilon_2)_{max}} = \left(1 + U_{12}^2(\omega_2) \frac{J_1}{J_2} + U_{21}(\omega_2) \cdot \frac{dU_{21}(\omega_2)}{d\omega_2} \frac{J_1}{J_2} \cdot \omega_2 \right)^{-1} \quad (3)$$

Simulated results are shown in figure 2:

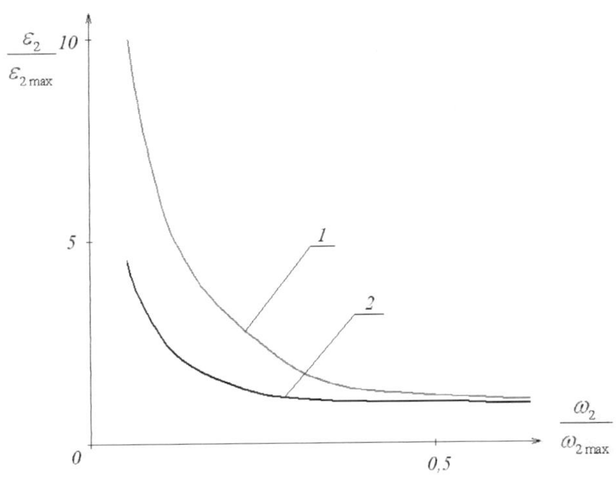

Figure 2. Dependence of the ratio of the current acceleration of the machine shaft to its maximum value on the ratio of the current speed to its maximum value for energy recovery (curve 1) and without energy recovery (curve 2).

Curve 1 shows that the acceleration of the vehicle ε_2 decreases as its speed ω_2 increases, this is typical for solving the first-degree linear equation, which is describing the transient acceleration of a transport vehicle with an internal combustion engine. The physical meaning of this reduction in acceleration towards the end of the movement when accelerating with energy recovery is that by the end of the movement there is a decrease in the kinetic energy of the flywheel and, of course, a decrease in the driving forces. Curve 2 describes the behavior of the same machine at a constant value of the transfer function, which corresponds to acceleration without energy recovery and the absence of a term with a derivative in the equation. Thus, the difference between the values ε_2 of curves 1 and 2 shows how much acceleration is reduced in the absence of mechanical energy from the flywheel battery operating in engine mode during acceleration. This is natural, since the total power of the main engine and the upper-stage flywheel engine in this case is lower.

A stand for conducting experimental studies was designed and manufactured to test the derived equations.

The scheme of the experimental modeling installation does not have a motor and speed sensors of the links. Measurement of the flywheel rotation angles was performed using a computer after high-speed photography of the movement of the mechanism. The flywheels of the four-ring mechanism exchange energy between themselves by braking one and accelerating the other. The experimental setup is started by setting one of the flywheels in motion. It is not possible to use an individual motor for this purpose, since the movement of the flywheels is reversible with a change in the direction of rotation after stopping.

In the experiment, the angular position of the flywheel was determined, and the gear ratio of the mechanism was determined by differentiating one angle of rotation of the flywheel from another angle, that is:

$$U_{12} = \frac{d\varphi_1}{d\varphi_2} \quad (4)$$

Then the law of changing the gear ratio by the angle of rotation of the rocker arm of the mechanism was obtained by calculation and experiment, was compared with the calculated gear ratio between the two rockers of the four-hinge mechanism.

Figure 3. Photo of the experimental stand mechanism.

The initial angle corresponding to the extreme (dead) position of the mechanism was assumed to be zero. On each frame, a line was drawn from the axis of rotation of the flywheel to the center of the crank attachment, from which the angle of rotation of the flywheel was counted. Stages of processing an experiment on a computer:

Figure 4. Initial position of the mechanism. A.

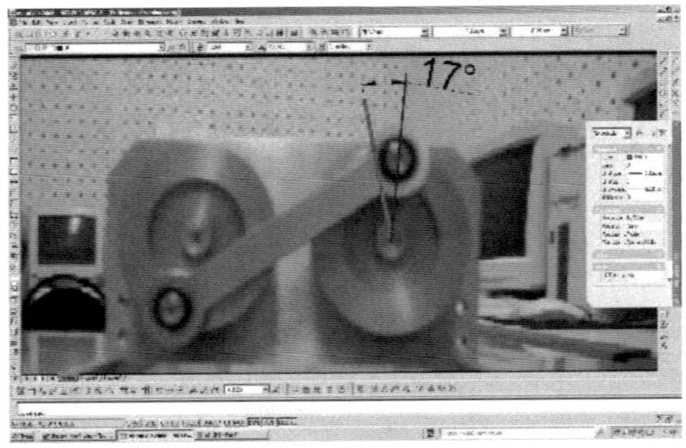

Figure 5. Initial position of the mechanism. B.

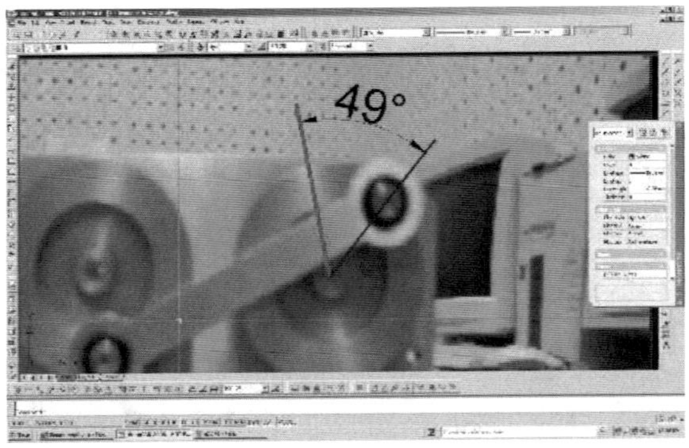

Figure 6. Initial position of the mechanism. C.

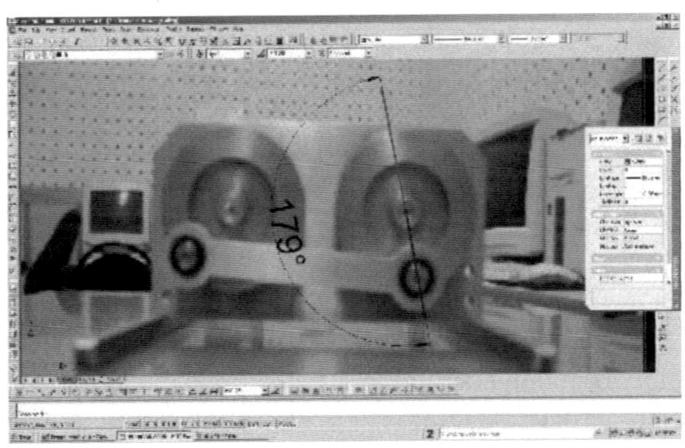

Figure 7. Initial position of the mechanism. D.

After that, the angles were measured in increments of 0.04 for each new position. To do this, a line was drawn in each subsequent frame connecting the flywheel rotation axis with the crank mounting axis. The angle between the zero position line and the line marking the disk position on each frame was then measured, as shown in figures B, C, D.

The obtained results of processing two cycles of the experiment on a computer were presented in the figure 8.

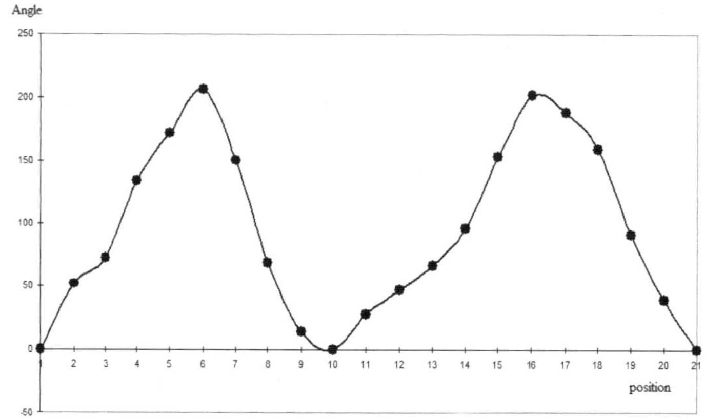

Figure 8. Example of experimental dependencies of the oscillations of the flywheel before carrying out the approximation.

Graphs of superimposed oscillations are plotted based on the obtained calculation and experimental data.

The deviation of the obtained curves from the theoretical data does not exceed 8%, which is an acceptable order of error [26]. Thus, it is possible to make a conclusion about the possibility of using a flywheel battery in hybrid cars to improve their environmental friendliness, this conclusion is confirmed by the experiment.

It should be noted that if it is necessary to construct a uniquely shaped flywheel according to the original design, it is possible to use additive technologies. Modern printers print at high speeds and are much cheaper than casting experimental flywheel molds in factories. In particular, by modeling a flywheel on a 3D printer, you can get the appearance and combinations of parameters that are relevant for a specific experiment on laboratory facilities [27].

3. Conclusions

1. A flywheel battery can reduce the acceleration time, as well as reduce the total operating time of the internal combustion engine, especially in places of forced stop at the time of driving in the city cycle mode.
2. The use of a hybrid power plant in a vehicle provides significant fuel savings and reduces environmentally harmful emissions.
3. Flywheel energy accumulators have a simple design, high resource, and their disposal causes less damage to the environment than the disposal of electrochemical batteries. 1. The study and calculation of the carbon footprint of hybrid solar-wind installations is an urgent task, the solution of which will help to understand the degree of influence of alternative energy on the environment.

Acknowledgments

The authors express their deep gratitude to Ph.D. Askarov Askar Fatovich for his invaluable support and constructive comments.

References

[1] Permyakova O G, Ovsyannikov V V, Kaminsky N S et al. 2019 Improving the Environmental Friendliness of Russian Transport Through the Use of Hybrid Vehicles *IOP Conference Series: Earth and Environmental Science* **459** 042079 DOI 10.1088/1755-1315/459/4/042079.

[2] Kiseleva E V, Kaminskiy N S, Presnykov V A 2020 Study of Fuel Efficiency of Hybrid Vehicles *IOP Conference Series: Earth and Environmental Science* **459** 022086 DOI 10.1088/1755-1315/459/2/022086.

[3] Feng L al. 2020 Optimal system parameters and hybrid ratio for fuel cell hybrid electric vehicles *Sensors and Materials* DOI 10.18494/SAM.2020.2694.
[4] Barbashov N N, Abdullina L R, Ilyushkov I N, Bolotov I E 2020 Research of influence of errors of cam manufacturing on the law of motion of output element *IOP Conference Series. Materials Science and Engineering*. **862** 032047 DOI:10.1088/1757-899X/862/3/032047
[5] Abdullina L, Barbashov N and Leonov I 2019 Analysis of Possibilities to Reduce Energy Consumption of Elevator Systems. In: Radionov A., Kravchenko O., Guzeev V., Rozhdestvenskiy Y. (eds) *Proceedings of the 4th International Conference on Industrial Engineering. ICIE 2018. Lecture Notes in Mechanical Engineering. Springer, Cham.* https://doi.org/10.1007/978-3-319-95630-5_18
[6] Zhang P *et al* 2015 A comprehensive analysis of energy management strategies for hybrid electric vehicles based on bibliometrics *Renewable and Sustainable Energy Reviews*. DOI 10.1016/j.rser.2015.03.093
[7] Weiss M, Zerfass A, Helmers E 2019 Fully electric and plug-in hybrid cars - An analysis of learning rates, user costs, and costs for mitigating CO2 and air pollutant emissions. *J Clean Prod*; doi:10.1016/j.jclepro.2018.12.019
[8] Coffman M, Bernstein P, Wee S 2017 Electric vehicles revisited: a review of factors that affect adoption. *Trans. Rev*, **37**.
[9] Asim Nadia, Adnan Al Bimani, Saqri Al Ahmed and Matriano Maria 2020 The Social and Environmental Impact of Hybrid Cars. *International Journal of Research in Entrepreneurship & Business Studies*, **1**. DOI:10.47259/ijrebs.114
[10] Santosh Hiremath 2020 Impacts of Hybrid vehicles on environment, *STM Journals,* DOI:10.37591/.v8i3.867
[11] Andrea Aurora Racz, Ionuţ Muntean, Sergiu-Dan Stan 2015 A Look into Electric/Hybrid Cars from an Ecological Perspective, *Procedia Technology*, **19**.
[12] Eckard Helmers, Patrick Marx 2012 Electric cars technical characteristics and environmental impacts, *Environmental Science Europe*.
[13] Vasant Kumar, Jikauan Yang, Seref Sonmez 2013 Relevance of reaction of lead compounds with carboxylic acid in lead recovery from secondary dources, *Cambridge*.
[14] Leila Abdullina, Vladislav Smirnov, Anna Alimova, Alina Kalistratova, Alexander Kravets 2021 Development of eco-friendly mechanized rotary parking lots with a flywheel energy storage device *IOP Conf. Ser.: Earth Environ*. Sci. **677** 052037.
[15] Chetvernin M U, Barbashov N N and Podolskiy A I 2021 Development of a method for diagnostics of the technical condition of spindles of metalworking machines *Journal of Physics: Conference Series, Intrumentation Technologies and Environmental Engineering*, **1889**.
[16] Zrazhevskiy A V, Kravtsov A G 2021 Existing hybrid power systems based on wind generators and prospects for their application, *Technical sciences: problems and solutions: collection of articles. Art. based on the materials of the LIV International Scientific and Practical Conference "Engineering Sciences: Problems and Solutions" Ed. Internauka* **11(49)**.
[17] Zrazhevskiy A V et all 2021 Improving the accuracy of designing a delta robot for 3D printing *Journal of Physics: Conference Series, Engineering and Materials Science*, **2094**.
[18] Zhu Tianjun et all 2017 Simulation Research on the Start-stop System of Hybrid Electric Vehicle *Journal of Advances in Vehicle Engineering.* **3**.
[19] Chai P 2015 *Design and Optimization of a Flywheel Based Kinetic Energy Recover System* (Graduate project, George W. Woodruff School of Mechanical Engineering), DOI:10.13140/RG.2.1.4095.6643
[20] Korsunskii V A 2013 Analysis of dynamic characteristics of a transport vehicle with combined power installation and hydrostatic transmission. *Nauka i obrazovanie: nauchnoe izdanie MGTU im. N.E. Baumana* **9** 145-62
[21] Barbashov N N 2012 *Improving the characteristics of flywheel energy batteries of the drive of*

hoisting-and-transport machines (Moscow)

[22] Pena-Alzola R, Sebastian R, Quesada J and Colmenar A 2011 Review of Flywheel based Energy Storage Systems. In Proceedings of the 2011 *Int. Conf. on Power Engineering, Energy and Electrical Drives (Malaga, Spain)*

[23] Zhu T et al 2017 Simulation Research on the Start-stop System of Hybrid Electric Vehicle. *Journal of Advances in Vehicle Engineering* **3(2)** 55-64

[24] Hoimoja H 2006 Flywheel energy storage: principles and possibilities *3rd Int. Symposium Topical problems of education in the field of electrical and power engineering (Kuressaare, Estonia)*

[25] A. Podolskiy et al. 2021 Calculation of the optimal power indicators for the engine of a small spacecraft *IOP Conference Series: Materials Science and Engineering* **899** 012012 DOI 10.1088/1757-899X/1155/1/012012.

[26] Barbashov N N, Samoilova M V, Abdullina L R 2021 Selection of rational algorithms for controlling high-precision details, *Journal of Physics: Conference Series, Intrumentation Technologies and Environmental Engineering,* **1889**.

[27] Zrazhevskiy A V 2021 Application of additive technologies in industry *Naukosphere* **8(1)**

AUTHOR INDEX